DESIGNING FOR SAFE USE

100 PRINCIPLES FOR MAKING PRODUCTS SAFER

DESIGNING FOR SAFE USE

100 PRINCIPLES FOR MAKING PRODUCTS SAFER

Michael Wiklund
Kimmy Ansems
Rachel Aronchick
Cory Costantino
Alix Dorfman
Brenda van Geel
Jonathan Kendler
Valerie Ng
Ruben Post
Jon Tilliss

CRC Press
Taylor & Francis Group
Boca Raton London New York

CRC Press is an imprint of the
Taylor & Francis Group, an **informa** business

Contents

An introduction to safety

What is safety?

Safety is the result of many factors, including the way people think and behave, characteristics of the environment, and protections against hazards that might be present. In some cases, safety is something that can be objectively measured. In other cases, safety might be subjectively judged, as something being safe or unsafe.

See *Principle 5 - Include pads*

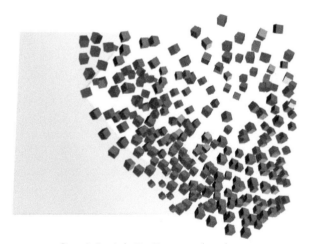

See *Principle 7 - Temper the glass*

Objectively speaking, a motorcycle helmet offers some protection to a rider in many accident scenarios. The trauma that results in the event that a rider's helmeted head strikes the pavement is bound to be less than if the rider was not wearing a helmet. The added protection can be demonstrated using strain gauges, accelerometers, and X-rays, for example.

Subjectively speaking, a rider is likely to feel safer when wearing a helmet, regardless of whether he or she understands or even cares about the product's specific protective properties (e.g., force vectoring, absorption characteristics). Such is the nature of protective equipment—to be there when it is needed, and otherwise fade into the background or be invisible.

For example, glass tabletops can be tempered so that, in the event that someone falls onto it, the glass breaks into lots of little bits, rather than dangerous shards. The fallen person need not be aware of the safety feature ahead of time to benefit from it when it matters most.

That said, people sometimes need to recognize hazards and take care to avoid them. This is the case when safety depends on someone reading a warning (e.g., "Keep hands away from moving gears") or donning a protective device (e.g., respirator). Therefore, safety is not always assured unless people participate in the process of staying safe.

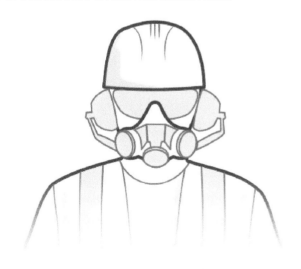

See *Principle 96 - Make PPE available and usable*

Ultimately, product designers have a responsibility to protect the people who come into contact with the product. Users assume that all types of gadgets, devices, tools, equipment, systems, and the like are reasonably safe to use as intended—"reasonable" being the keyword.

But what does "reasonably safe" mean? What is the safety threshold between reasonably safe and not? This is for society and individuals to decide, and the decisions are closely tied to risk perception. If the goal were to achieve absolute safety, there would be no motorcycles, or cars for that matter. There would be no contact sports. There would be no knives with which to slice bagels in half. But, we routinely accept and take risks, including walking down a street that has manholes that we expect to be covered, but might not be.

The hypothesis of risk homeostasis[1] holds that people establish a threshold for themselves with respect to the amount of risk they will accept. This explains why an average skier might fly down the beginner's slope, yet take his or her time on the expert slope, therein accepting about the same risk of an injurious fall. Accordingly, we (society as a whole) tend to invest resources into making things safe enough that users are mentally comfortable with their risk exposure. In some cases, and due to personal standards of care and social norms, we accept considerable risk (e.g., riding a motorcycle), and in other cases, very little risk (e.g., undergoing LASIK eye surgery).

One more dimension of safety to consider is exposure. You assume no risk of head trauma due to a motorcycle accident if you never ride a motorcycle, excepting the case that one hits you. Likewise, you assume no risk of hand amputation due to operating a metal brake (i.e., a bending machine) if you never operate one. Therefore, individuals are already quite safe in view of all the hazards in the world. But, it is not always possible to control when an exposure might occur. Similarly, we cannot always count on people to have sufficient familiarity with hazards and proper precautions to be safe when unexpected exposures do occur. Therefore, designers have the sobering responsibility to make things as safe as practical and, in some cases, as safe as absolutely possible. Notably, the latter option requires what could be significant and essential financial investments (e.g., producing a table saw that includes a mechanism to instantly stop a saw blade if it comes into contact with one's skin). When safety features help bring risk into line with most users' acceptance thresholds, products may enjoy commercial success. Consider the rewards associated with designing for safe use: products can command a higher price, manufacturers can give themselves greater liability protection, and designers can feel good about their work.

Words of encouragement

We encourage you to draw upon this book's principles for safe design to produce—yes—safer products. We also encourage you to draw guidance from other sources (particularly safety standards), use good professional judgment, and follow an iterative user-centered design process that includes user testing (also called usability testing).

1. "About Risk Homeostasis: A Theory about Risk Taking Behaviour." Gerald J.S. Wilde Ph.D., Professor Emeritus of Psychology, Queen's University, Kingston, Ontario, Canada, http://riskhomeostasis.org.

About the book chapters

We crafted each of the "chapters" conveying design principles with the following criteria in mind:

- Stay focused on principles that pertain to how users interact with products and how users can be protected against harm.

- Provide some specific user interface design tips.

- Illuminate an aspect of designing safe products that either has broad application, is particularly important to certain types of products, and/or is simply interesting.

- Use photos of real products to illustrate some of our points.

- Make the chapter something we would like to read.

- Include important details that are central to understanding how to achieve safety with regard to the aspect of safety in discussion.

- Use our professional judgment and a synthesis of available information on a topic to make it more accessible to non-specialists.

- Include some "fun facts."

- Be artful.

- Add humor—primarily via graphical depictions—to an otherwise serious subject for the sake of reading enjoyment.

Each chapter is accompanied by an icon, meant to make each principle perhaps just a bit more memorable. On the next page is a small assortment of these icons.

As you delve into the chapters, you might notice that there was a lot to say in words about some topics, while other points were best communicated in a principally graphical manner. So, you will see stylistic variety among the chapters, and you will notice that some chapters are faster reads than others.

We acknowledge and applaud the fact that people with greater subject matter expertise could probably write a book on each of the 100 topics. But, we are pleased with our final product. We hope you enjoy the book and take away some important lessons.

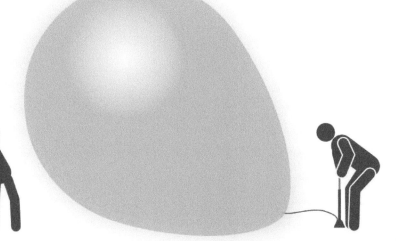

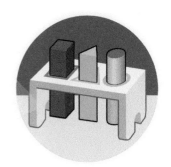

About the exemplars

As "bonus content," we have complemented the safe design principles with 10 exemplars of designing for safe use. Each exemplar is a showcase—an amalgam of safety features that we found in actual products. However, our renderings of hypothetical products are not intended to represent any actual products on the whole.

Exemplar 1
Car seat

Exemplar 2
Diabetes Management
Software

Exemplar 3
Tractor

Exemplar 4
Stepladder

Exemplar 5
Anesthesia
machine

Exemplar 6
Chainsaw

Exemplar 8
Steam iron

Exemplar 7
Medication blister pack

Exemplar 9
AED

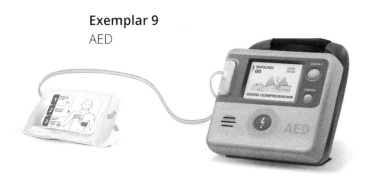

Exemplar 10
Stretcher

IMPORTANT: We are not endorsing actual products that have one or more of the cited safety features. A particular product's safety must be judged holistically and in view of test data, such as what might be collected through tests involving the given product and its intended users. Even then, there is a need for post-market surveillance to determine if a thoroughly tested product actually performs safely in the real world.

About the endnotes

The endnotes section, which you will find at the back of the book, lists the sources of quoted material and related content that you might wish to access directly. We tried to stay on the correct side of the fine line between referencing sources for uncommon and highly specialized information versus what many people would consider to be common knowledge.

This section also presents credits to the individuals and organizations that generously gave us permission to use their photographs. All other photos were our own or did not require permission.

In conclusion...

A message from SAM (Safety Action Model)

I know what you're thinking: that I look really familiar. That's because you've seen me in lots of warning signs and instruction sheets, giving you a heads-up when there is a hazard in the midst and you need to be cautious.

Some of my classic appearances include:

HIGH VOLTAGE!   CRUSH!   WET FLOOR!

I show up on many pages throughout this book, helping to identify hazards, which inform principles of designing for safe use. It's my job and I love what I do.

So, don't worry about me. I'm a professional model—technically an Isotype[2]—and it's my job to be exposed to hazards and graphically suffer injuries so that people can stay safe.

My hope, and the hope of the authors of this book, is that our content helps drive efforts to make products as safe as possible so that real people do not suffer the type of real-life consequences that I only experience on paper.

2. "The History of Symbols." Graphic Design History, Designhistory.org, 2012, www.designhistory.org/Symbols_pages/isotype. html.

Acknowledgments

We have numerous colleagues, teachers, friends, family, and others to thank for supporting our book project. Their support took many forms, such as guiding us to be creative, sharing ideas on how to depict safe and unsafe scenarios, commenting on draft content, delivering food and beverages to our workstations, enduring the many hours that we spent on the project rather than socializing with them, and inspiring us to do our best.

Provide stabilization

Principle

Prevent products from tipping over or falling on someone by adding stabilizing features or securing them to their surroundings.

Know your center of gravity

Toppling objects can cause injury as well as property damage. Consider the consequences of a parked motorcycle falling onto a bystander, a ladder falling sideways with a house painter aboard, and a chest of drawers pitching forward onto a toddler who is climbing it to fetch a toy (see *Principle 63 - Childproof hazardous items*). An object is naturally stable when its center of gravity is within its footprint rather than outside of it. Move the center of gravity beyond the footprint (i.e., beyond the area of support) and the result is instability plus the potential to tip over. It also helps to keep the center of gravity low so that applied forces, such as the centrifugal force acting on a riding lawnmower turning sharply, do not overcome the stabilizing forces.

Better

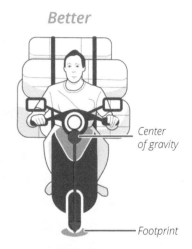

Center of gravity

Footprint

The motorcycle is stable because the center of gravity is within the motorcycle's footprint.

Tip-overs are no joke

Between 2000 and 2013, the US Consumer Product Safety Commission reported that product tip-overs and/or instability led to 430 deaths. They also reported that tip-overs caused an average of 38,000 injuries requiring emergency department visits per year during the period of 2011 to 2013. Most of those injuries (56%) were caused by furniture tip-overs. The most common injuries were contusions, abrasions, and internal organ damage. The most commonly injured body part was the head, followed by the legs, feet, and toes. Children between the ages of 1 month and 10 years accounted for 84% of the reported deaths.[1]

Worse

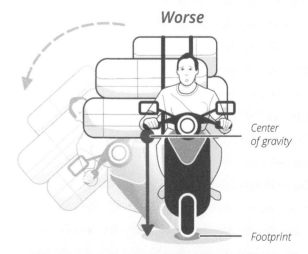

Center of gravity

Footprint

The motorcycle is unstable and prone to tipping over because the center of gravity is outside the motorcycle's footprint.

Never go beyond your footprint

When you add a person into the mix—such as one standing on a stepladder—the key is to keep the person reasonably centered within the footprint, rather than extending past it. In the case of stepladders, railings (see *Principle 32 - Provide a handrail*) help to do this as does making the steps narrower as the height increases. A crossbar-type of hand hold also encourages safe positioning.

Training wheels increase a bicycle's footprint, which helps children keep their balance.

Stability improves as you enlarge an object's footprint without changing other physical variables (e.g., height, weight). That is why an office chair with a five-leg base is inherently more stable than one with a four-leg base.

To prevent tip-overs and still handle heavy loads, lift trucks and cranes have outriggers that extend to increase their footprints so that they remain under the entire apparatus's center of gravity. Outriggers of a sort are also used to stabilize the top of some roofing ladders.

Cranes use extensions (outriggers) to increase the crane's overall footprint and create added stability.

If you cannot keep an item's center of gravity within its footprint, or if forces might overturn the object, then some kind of retention or hold-down device might help. Common examples are brackets and tethers used to keep ovens and chests of drawers/ bookshelves from toppling forward.

Make things easy to clean

Principle
Reusable products should be designed to facilitate easy and effective cleaning, sterilization, and/or disinfection.

Clean up nicely

Perhaps the biggest risk associated with things becoming dirty is exposure to chemical and/or biological agents. That's why the food and medical industries have standards for cleanliness and even sterility.

Aside from causing contamination, filth can make labels and warnings illegible, make surfaces slippery when wet, and interfere with mechanical motion. For example, consider the following scenarios:

- A poorly located warning label could be obscured over time by soot from a machine's exhaust pipe or by clippings from a mower's discharge chute.

- A greasy handle could lead someone to lose his or her grip while lifting a heavy object.

- Gears contaminated by dirt could fail to mesh properly, possibly leading to a dramatic mechanical failure or perhaps a motor overheating.

Accordingly, manufacturers of reusable devices should ensure that devices facilitate easy and effective cleaning, as well as any necessary disinfection or sterilization.

Toppings no one ordered

Easy cleaning should also be a priority in the food industry. Restaurants and establishments serving a high volume of customers typically use several devices to help expedite food prep in the kitchen. Unfortunately, some of these devices can have nooks and crannies that harbor guck and unwanted bacteria.

In July 2017, a McDonald's employee was fired after he tweeted pictures of an ice cream machine drip tray filled with mold (see image on left).[1]

Ways to make things easy to clean

FDA and reprocessing

Sobering to those who have undergone or will undergo a related examination, the US Food and Drug Administration (FDA) recently compelled manufacturers of minimally invasive devices, such as duodenoscopes, to develop designs and cleaning procedures to prevent cross-patient contamination.

After a series of illnesses—and in a few cases, deaths—occurred due to accumulated bacteria, the FDA ordered three duodenoscope manufacturers in the US (Olympus, Fujifilm, Pentax) in 2015 to conduct postmarket surveillance studies so the FDA could better understand how duodenoscopes are reprocessed in real-world settings.[2] In a statement to *USA TODAY*, the FDA said it is studying the problem and collaborating with manufacturers to determine whether the cleaning protocols can be revised, or whether obstacles to thorough cleaning require an entire redesign of the scopes themselves. Meanwhile, it is "important for these devices to remain available" because of their "lifesaving" ability to detect and treat potentially fatal digestive disorders.[3]

Based on the duodenoscope-related incidents and studies described above, the FDA updated their guidance on reprocessing medical devices in healthcare settings in 2017.[4]

Eliminate or minimize seams, crevasses, and places for fluids and materials to collect.

Ensure that people can reach areas that require cleaning, and that it is easy to find and access areas that require cleaning.

Use material finishes (e.g., lacquer, plastic coating, anodizing) that enable thorough cleaning without degradation.

Use a color that will make grime visible, highlighting when a device requires cleaning.

Eliminate or minimize the need for special cleaning apparatus, materials, and solutions, but make them readily available if they are necessary.

Eliminate or minimize the need for item disassembly and reassembly in order to clean it.

Convey proper cleaning procedures via instructions, a quick reference guide, a checklist, and/or on-product labels.

Eliminate small parts from kids' products

Principle

When designing products that will be used near or by children, eliminate small, loose parts or objects that could cause choking.

The Small Parts Regulations

Anyone who has small children or who has been around small children knows that children will put almost anything in their mouths. But, loose, small parts are a hazard to young children. That was the evidence-based conclusion drawn by the US Government that led to the Small Parts Regulations - Toys and Products Intended for Use by Children Under 3 Years Old, 16 CFR Part 1501 and 1500.50-53, which is enforced by the US Consumer Product Safety Commission (CPSC).[1]

CPSC states, "This regulation prevents deaths and injuries to children under three from choking on, inhaling, or swallowing small objects that they may 'mouth.' It bans toys and other articles that are intended for use by children under three and that are or have small parts, or that produce small parts when broken."

How do I know if a part is too small?

A part is too small for incorporation into a product intended for young children if it fits completely into a specially-designed test cylinder specified by the CPSC that is often referred to as a "choke test cylinder"[2] (shown on the right). The cylinder is a simple model for a 3-year-old child's expanded throat. If an item fits into the cylinder and is not on a list of exempt items per CPSC (e.g., buttons, chalk, barrettes), it is considered a choking hazard and banned because it could possibly fit down a young child's throat and create a blockage.

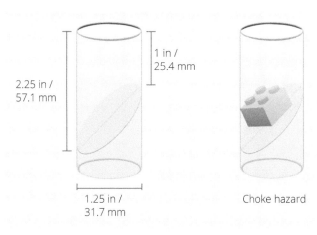

2.25 in / 57.1 mm

1 in / 25.4 mm

1.25 in / 31.7 mm

Choke hazard

The choke test cylinder models a 3-year-old child's expanded throat (left). If an item fits completely in the cylinder, it is considered a choking hazard (right).

Keeping up with the big kids

The challenging fact is that children under the age of 3 often play in the same space as older children who have playthings with small parts that are not subject to the CPSC regulation. Also, there are likely to be plenty of small objects in the environment (e.g., coins, keys, pen caps) that are not toys and could also pose a choking hazard. Still, the regulation helps protect young children from the parts of the products intended explicitly for their use.

This symbol must be added to CE-marked toys that contain small parts that are unsuitable for children less than 3 years.[3]

So, manufacturers of young children's products have their "marching orders." No product shall, as an integrated whole or as a separated portion thereof, fit entirely into the simple and clever choke test cylinder. Meanwhile, those designing and manufacturing products for older children and adults should be mindful that their products could get into the hands and mouths of little ones. Therefore, manufacturers should avoid producing unnecessarily small parts that are loose or might become loose. In general, manufacturers should take care to "childproof" products (see *Principle 63 – Childproof hazardous items*).

Here are some ways to prevent small parts from becoming a hazard:

- Permanently tether small parts, like a gas cap to a car's filter port or pens to the desk at a bank.

- Make small parts slightly larger than absolutely necessary so that they do not fit in the choke test cylinder.

- As might be required already, place a warning on the product package to indicate that a plaything intended for older children, for example, is not intended for use by children 0-3 years of age (see symbol above).

Quick Tip

If you are looking for a home-test and do not have a choke test cylinder, some suggest using an empty toilet paper roll. Its diameter is slightly larger than the choke test cylinder, enabling you to err on the side of caution.[4]

Examples of choking hazards

Hot dogs
A hot dog becomes a serious choking hazard to young children because it is "just the right size and consistency to perfectly block the airway."[5]

Marbles
Marbles intended to be enjoyed by older children are a choking hazard to young children.[1]

Small objects
Small objects (less than 1.5" in diameter) are reported as particularly hazardous. Such objects include buttons, coins, balls, and building blocks.[6]

Fruit
Cherry tomatoes and grapes can be a hazard, which is why they should be quartered before feeding them to young children (and yes, tomatoes are a fruit).[7]

Limit sound volume

Principle

If a product can generate loud sounds ≥ 85 decibels,[1] which can cause temporary or permanent hearing loss or cause severe discomfort, it should incorporate volume-limiting features.

Can you hear me?

Generally speaking (or perhaps shouting), loud sounds are bad for ears. Just as looking into the sun too long can damage one's eyes (causing a condition called solar retinopathy),[2] being exposed to a very loud sound, or even moderately loud sound continuously, can damage the sensitive structures in the inner ear—sometimes permanently. The damage is termed Noise-Induced Hearing Loss (NIHL).[3] This explains why some artillery corps veterans have hearing loss. It also explains why folks who work around loud machinery are encouraged, or even required, to wear hearing protection. Not only can loud sounds lead to hearing loss, loud sounds can also arguably be annoying and disrupt communication.

What generates loud sounds?

Components striking or rubbing against each other, vibrating components, pressurized gas escaping through a nozzle, and cooling fans commonly cause loud sounds that can damage hearing over time.

The US Occupational Safety and Health Administration (OSHA) has established guidelines for occupational noise exposure. The PEL (Permissible Exposure Limit) is based on noise level and exposure time. According to OSHA, hearing damage can occur due to exposure to an 90 dBA sound (e.g., the level of a power lawnmower) for 8 hours per day. The standard also indicates that exposure time should be reduced by half with each noise level increase of 5 dBA.[4,5] So, even being exposed to a very loud sound for a short period of time can cause hearing damage. For example, a nearby lightning strike producing a clap of thunder registering 120 dB can cause permanent hearing loss.[6]

Permissible Noise Exposures. When users are exposed to sound levels that exceed the levels presented above, and sound-reducing controls cannot be utilized, users should make use of personal protective equipment to reduce sound levels to permissible sound levels.[5]

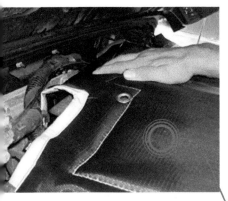

Protection against loud sounds

Wearing Personal Protective Equipment (PPE), such as ear plugs and ear muffs, can protect wearers against hearing loss. So can moving far away from the source. Product developers can also help by designing products to be quieter, either by lowering the amount of sound energy a product generates, or by muffling it. Here are some options:

- Reduce collision forces (that would otherwise produce a very loud sound)[7]

- Add padding that absorbs sound (i.e., passive noise cancellation)[7]

- Dampen the sound with a counteracting waveform (i.e., active noise cancellation)[8]

- Emit sound away from the user

When these sound energy reduction methods are not practical, you can:

- Encase the sound emitter (i.e., create a barrier to prevent sound escaping

- Add a muffler (i.e., silencer) that absorbs some of the sound energy, as in the case of an automobile with an internal combustion engine emitting high-pressure exhaust gases[9]

If it is impossible to reduce the product's sound to an acceptable level, warn users about the noise-related hazard, direct them to wear PPE (see *Principle 96 - Make PPE available*), and indicate the potential harm if they do not heed the warning.

Note that making a product too quiet can also be dangerous. For example, the sound of an engine helps pedestrians detect an approaching vehicle. Therefore, electric or hybrid cars that are nearly silent are sometimes difficult to detect, especially if a pedestrian is visually impaired. It is 40% more likely for a pedestrian to be struck by a "silent" car than by one with a noisy engine.[10]

Noise cancellation

Noise-canceling headphones are a great tool for reducing the consequences of constant exposure to loud sounds, such as those experienced by private pilots in single engine aircraft. Noise-canceling headphones work by using a microphone to detect ambient sounds, and then using electronics in the ear piece to generate noise-canceling waves that are 180° out of phase with the ambient noise waves, effectively "erasing" the ambient sounds.[8] Manufacturers now also apply noise-canceling technology that reduces noise at the source, eliminating the need for people to wear noise-cancelling headphones or hearing protection.

Include pads

Principle
Add padding to protect users from forceful and/or repeated impacts.

Pain, bruises, sprains, and broken bones be gone!
Padding reduces the potential for immediate and cumulative trauma. At least that is the idea behind incorporating padding into many products. However, the padding must be designed properly to ensure it delivers the intended level of protection.

A motorcycle helmet's hard shell is an important defense against the superficial effects of blunt trauma such as abrasions and lacerations. However, it is the helmet's internal padding that diminishes the peak forces applied to the brain—forces that could otherwise cause more severe and lasting injuries. The internal padding extends the amount of time the head (and the brain within) has to accelerate or decelerate depending on the impact scenario, with milliseconds sometimes making the difference between lesser versus greater harm.[1] The idea is to reduce peak forces by dissipating them over a wider area and longer period of time.

Impact force with and without padding

Impact energy *(y-axis)*

Without padding

With padding

Time

Padding slows down the impact force, spreading the force over a longer period of time and reducing the peak force.[1]

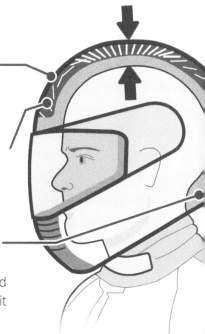

Rigid outer shell
Protects against abrasions and lacerations

Impact-absorbing foam liner
Absorbs most of the impact/force

Interior padding
Absorbs the remaining force and provides a secure fit

Multiple layers inside a motorcycle helmet work together to absorb force and protect the rider's head.[2]

Pads in everyday life

Bicycle seats are padded for comfort's sake, but also to reduce pressure on male and female anatomy "down there" that can lead to pain and various physical dysfunctions. Crutches are padded for comfort and safety as well. Padding at the top protects the underarms—though users should not rest their sensitive underarms on their crutches—and the padded grip spreads out the force on the user's hands while also ensuring a secure grip.

Some of the safer trampolines have padded perimeters to protect jumpers (typically children) from direct impact with the product's frame, springs, and other hard or sharp components. The padding around basketball backboards serves the same purpose, protecting people with impressive jumping ability if they hit their head on the backboard.

Shock-absorbing materials within automobiles serve to protect occupants during an accident. Preston Tucker is credited with designing the first automobile featuring a padded dashboard—the 1948 Tucker Sedan (also known as the "Tucker Torpedo")—intended to make the vehicle safer in a frontal collision.[3]

How to make padding more effective

Below are some recommendations for making padding more effective:

 Make padding thick enough that it will not bottom out.

 Enable users to remove pads for cleaning or to replace them if they are damaged.

 Design padding to stay put rather than move and potentially uncover areas where users are left unprotected. Snaps and Velcro™ are often used for this purpose.

 Use padding materials that will withstand environmental factors such as light, moisture, and vibration.

 Use padding materials that will withstand routine wear and tear.

 Use fire-resistant materials that will not catch fire.

Make it buoyant

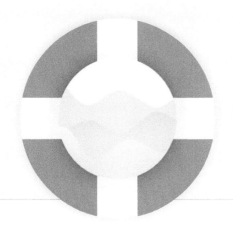

Principle

If a product will be used in situations where users could drown or suffocate, make the product buoyant to help users float and survive.

Floating around

There are probably few canoes that have not tipped over at some point. Fortunately, most canoes are designed to float, giving the newly submerged swimmers something to hold onto and possibly get back into. A wooden canoe's buoyancy might stem from its materials or construction, but also from its built-in air chambers or supplemental air bags. After all, there are few things that enable flotation as much as plain old air. For example, small pockets of air in Styrofoam™ make most Styrofoam objects buoyant.[1]

Water wings

Water wings were invented to help people—primarily children—float in water that might be over their heads. However, water wings cannot prevent drowning if they slide off the wearer.[2] Keeping flotation devices on requires a good fit and a guardian's vigilance, which means that flotation devices do not fully eliminate the risk of drowning. A zipped life jacket is likely to be a safer solution, with water wings serving a transitional role along the path toward swimmer competency and independence from flotation aids.

Airbags

One safeguard against perishing in an avalanche is wearing an airbag that can be inflated in an emergency. The device, which looks similar to other flotation devices, helps to keep the victim atop the roiling snow pack because larger and less dense objects tend to rise or "float" to the top, and smaller objects tend to sink.[3]

Flotation suits and life jackets

Mariners don special suits to survive disasters at sea. Such suits are designed to keep a wearer's head above water even when she or he is unconscious. A good life jacket is supposed to function in the same manner.[4] The key is not only to keep users afloat, but also in the proper orientation—obviously with their head up.

Aircraft life jackets

Aircraft life jackets are designed to be compact, which is why they look different from the life jackets usually stowed on boats. They incorporate compressed CO_2 canisters that, when triggered, inflate the vest. As an added safety feature accounting for canister failure, they also include breath tubes that enable manual inflation.

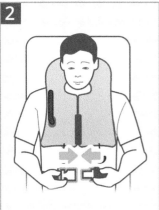

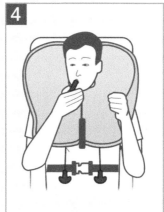

Keep floating

Ideally, items designed to float should not lose their buoyancy due to aging, wear and tear, or even significant damage. For example, a life jacket should not wear out after a few seasons. Similarly, air-filled bladders should not be vulnerable to popping, or there should be enough of them so that damage can be isolated to one pocket and avoid putting the wearer in jeopardy. In keeping with this principle, inflatable rafts that traverse rocky rapids have multiple, independent air chambers.

Make it buoyant

The following tests help ensure that flotation devices are effective:[5]

Donning test
When using a flotation device in an emergency, the user should be able to don the device in 60 seconds.

Turning test
The device should turn an unconscious wearer to face upwards or enable a conscious wearer to turn to face upwards.

Flotation stability test
The user should not need to perform additional actions to stay afloat.

Water entry test
The device should not slide off or injure the user when the user enters the water.

Temper the glass

Principle
Ensure glass is tempered to prevent it from shattering into dangerous shards that could cause injury.

Advantages of tempered glass

Standard (i.e., annealed) glass can shatter into dangerous shards, whereas stronger, tempered glass, if subjected to enough destructive force, will disintegrate into small pieces that are less hazardous. Therefore, tempered glass, also known as toughened glass, or a type of safety glass, is used in applications where strength is needed and glass shards could injure people (e.g., in a glass elevator like the one on the right).

Automobile side and rear windows are usually tempered for these reasons. Their safety results from their ability to shatter into less dangerous pieces, as well as the ease with which they can be intentionally shattered with a center punch tool to enable escape from a wreck.

Unlike side and rear windows, automobile windshields are typically made of laminated glass (two layers of glass bonded together by a heat-adhered layer between them). That is why windshields hold together (rather than breaking into tiny pieces) when shattered, helping to keep people inside the vehicle safe in the event of an accident.

Some, but not all, glass tables are also tempered for strength and safety's sake, noting that falling onto and breaking a tempered glass tabletop is less likely to cause injury than falling onto an annealed glass tabletop, which can break into shards that can cause major wounds.

How does tempered glass receive its strength?

Tempered glass can be created from annealed glass through a process of controlled thermal treatments that strengthen it. The glass is heated in a furnace, up to 620 degrees Celsius or 1,150 degrees Fahrenheit, and then rapidly cooled to room temperature. Because the glass cools much faster at its surface than at its center, the outer surface compresses, while the center expands. The resulting compressive forces from the outside of the glass push against the expanding forces on the inside. The resulting tensile stress gives the tempered glass its strength, which can be 5 to 10 times stronger than annealed glass.[1]

Tempered Glass
The process of tempering glass results in strong glass that crumbles into small granular chunks, which are less likely to cause injury than annealed glass shards.

Annealed Glass
Annealed glass is commonly used in windows and breaks relatively easily, producing long, sharp splinters when it breaks.

Laminated Glass
Laminated glass will crack under pressure, but remain intact as the multiple vinyl layers hold its pieces in place and minimize the risk of cuts.

The 200-pound gorilla in the room

A broken smartphone screen might seem like a mere inconvenience, but it could actually be hazardous in certain situations. For example, a broken touchscreen might cut a user's hands or face or leave a stranded accident survivor with an inoperable phone. That's why some smartphones are made with impact-resistant glass, such as Corning® Gorilla® Glass. Gorilla Glass is toughened through a process in which sodium ions in the glass are replaced with larger potassium ions to create "compressive stress."[2]

Glass safety standards

ANSI Z97.1-2015 (Safety Glazing Materials Used In Buildings - Safety Performance Specifications and Methods of Test) provides the minimum requirements for glass safety. The standard "establishes the specifications and methods of test for the safety properties of safety glazing materials (glazing materials designed to promote safety and reduce the likelihood of cutting and piercing injuries when the glazing materials are broken by human contact) as used for all building and architectural purposes."[3] Shower doors are a good example of a product that benefits from this standard, given that a slip in the shower stall could cause someone to fall heavily against the door and potentially break it. No need to re-create a bloody spectacle reminiscent of the infamous shower scene in Alfred Hitchcock's *Psycho*.

Provide tactile feedback

Principle
Provide tactile feedback such as vibration, resistance, and texture when visual and/or audible feedback is impossible or ineffective.

Action and reaction

Tactile feedback can be a particularly effective way to communicate information to people while they perform tasks, especially when other communication means (e.g., visual and audible feedback) are unavailable or when senses other than touch (e.g., sense of sound or sight) are overloaded.

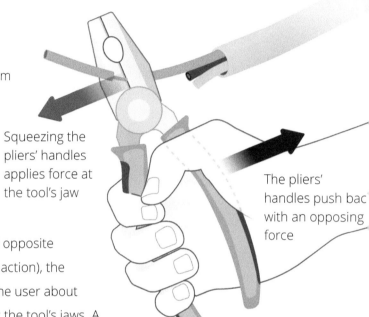

Squeezing the pliers' handles applies force at the tool's jaw

The pliers' handles push back with an opposing force

Tactile feedback relates directly to Newton's Third Law of Motion: For every action, there is an equal and opposite reaction. When a user squeezes a pliers' handles (the action), the handles push back (the reaction), thereby informing the user about his/her grip on the tool and the force being applied at the tool's jaws. A surgeon is likely to depend on a scalpel's shape and the associated feel to know which way its cutting edge is pointing (see *Principle 82 - Add shape-coding*). Or, a surgeon might need to sense button movement to confirm that she has activated a cranial drill.

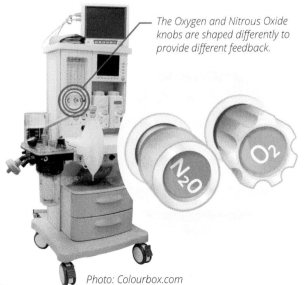

The Oxygen and Nitrous Oxide knobs are shaped differently to provide different feedback.

Photo: Colourbox.com

An anesthesiologist needs to sense different knob shapes to distinguish the oxygen control knob from the nitrous oxide control knob. The tactile feedback enables him to control the anesthesia machine without having to look at the knobs.

In many cases, it helps to supplement one type of feedback with another to help ensure good communication. This is the case when you press an elevator call button and it lights up, thereby providing both tactile and visual feedback.

Losing touch, losing control

Insufficient tactile feedback can cause dangerous uncertainty. For example, many of the automatic transmission selectors in today's automobiles do not move the way they used to move. They return to the same position after momentarily being moved forward or backward. This behavior eliminates the tactile feedback of moving the selector a greater distance to a definitive position that can help the driver to identify the selected mode (e.g., reverse rather than park). Many drivers suffered injuries—and, in one case death—because they mistakenly believed their cars were in park.[1] As such, newer automatic transmission selectors are not as good communicators as older ones.

Worse   *Better*

It used to be that airplane yokes shook violently at the onset of a stall—the point at which an airplane loses lift, potentially because it moves through the air too slowly or because the wing's angle of attack is too severe. But, as aircrafts became more advanced and control systems insulated pilots from signs of a stall, pilots lost the tactile feedback necessary to rapidly recognize and correct a stall. This led aircraft designers to add "stick shakers," which produce an artificial shaking sensation when a stall is detected, thereby providing pilots with the tactile feedback that they need to take action.[2]

"Stick shakers" cause the aircraft's yoke to artificially shake in certain conditions, thereby providing tactile feedback to the pilot.

Other examples of tactile feedback

Tactile Ground Surface Indicators (TGSI) are used to assist vision-impaired pedestrians.[3] The textures are noticeable by a long cane or by foot, and warn pedestrians of approaching streets or hazardous situations (e.g., the edge of a train platform) (see *Principle 88 - Provide Guidelines*).

In many industries, touchscreens are replacing hardware-based user interfaces. However, touchscreens have a significant disadvantage—they do not provide tactile feedback to the user, which can lead to use errors. Some touchscreen-based devices have addressed this shortcoming by providing haptic feedback—a way of simulating tactile sensations through vibrations or other motions—to inform the user that the input (e.g., a button press) was received. For example, some smartphones have a "vibrate on tap" feature that generates a small vibratory pulse when users press certain on-screen controls.

Make it fire resistant

Principle
Products that could be exposed to heat or flames should be made of fireproof, fire-resistant, or self-extinguishing materials.

Fire-resistant materials

Fires can spread quickly, doubling their size in seconds and raging out of control in less than 30 seconds.[1] Given such risks, products should self-extinguish if they catch fire or prevent a fire from happening in the first place.

That is why many home furnishings, including upholstered furniture, mattresses, and curtains, are made of fabric and foams that have been chemically treated with a fire retardant. That is also why, in the US and many other countries, children's sleepwear must be flame resistant and, if it does catch on fire due to direct exposure to an intense heat source, self-extinguish.[2] However, health concerns regarding the use of fire-retardant chemicals is growing—they have been linked to cancers, memory loss, lower I.Q.s, and impaired motor skills in children.[3] Such concerns have led some manufacturers to use naturally flame-resistant materials such as wool in place of other chemically-treated materials (see *Principle 10: Use non-toxic materials*).

Nomex®
Special-purpose fabrics such as the well-known Nomex, which was first marketed in 1967,[4] protect race car drivers and firefighters from exposure to high temperatures and flames. These materials essentially work as shields. They self-extinguish and are poor heat conductors, which is a good thing. These traits give the wearer valuable seconds to escape from searing heat and flames, such as those that might engulf a race car after an accident.

Fire blankets

In extreme situations, such as a forest fire, firefighters can dig into the ground and cover themselves with a fire blanket. The foil blanket can help them survive for some time as the fire burns above them.

A new type of fire blanket that promises even greater protection is made from a ceramic fiber coated in a zirconia/inconel spray, similar to the material used by NASA to protect space vehicles during their fiery re-entry to earth's atmosphere.[5]

Fire- and heat-resistant gloves

Oven mitts can be made of the aforementioned Nomex, guarding against burns and potential fire due to contact with a heating element or open-flame burner. Similarly, tasks such as placing logs into fireplaces, blowing glass, and welding are facilitated by heat- and fire-resistant gloves.

Self-extinguishing batteries

As new technology develops, our lives are filled with more and more electrical devices. Many of these devices (e.g., smartphones, laptops, tablets) contain lithium-ion batteries, which can be volatile and catch fire or explode if their enclosing device overheats.

In fact, electrical failures and malfunctions have become significant causes of fires in homes and other environments. In 2017, several Samsung Galaxy Note 7 batteries caught fire and/or exploded due to manufacturing problems, including the use of irregularly sized batteries that caused overheating.[6] Such problems have led researchers at Stanford University to develop a self-extinguishing lithium-ion battery.[7]

Fire testing

Many countries require products to be tested for fire safety before they can be marketed. Such testing can evaluate if a product is likely to start a fire, how it reacts to fire, and how effectively it self-extinguishes if it catches on fire.

In the United States, Underwriters Laboratories (UL) has been conducting fire testing for well over a century. UL analyzes field data and conducts live burn experiments to examine fire behavior, and has developed computational modeling techniques to better evaluate potential fire risks scientifically.[8]

Use non-toxic materials

Principle

Ensure that materials used in products are non-toxic, and will not leach, outgas, or otherwise break down in a harmful manner over the life span of the product.

Learning as we go

There is a litany of products that were once considered safe, but are now considered dangerous. Telephone poles were coated with creosote, penta, and arsenicals—materials now linked to a wide array of health problems including cancer and birth defects.[1] Bakelite was an early plastic that contained asbestos fillers, which were handled openly and potentially inhaled before workers understood the health risks.[2] More recently, the compound bisphenol-A (commonly known as BPA), which has been used to strengthen plastics and could be found in bottles and many more consumer products, has been a cause for concern. The material has been suspected of interfering with proper hormone function.[3] The risk led the FDA to ban the sale of baby bottles containing BPA back in 2012.[4]

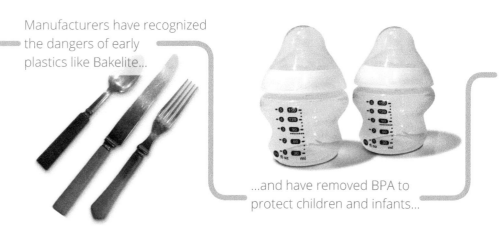

Manufacturers have recognized the dangers of early plastics like Bakelite...

...and have removed BPA to protect children and infants...

...whereas other manufacturers have starting using alternative materials that create a "green" look.

On the occasion that a material presumed to be safe is actually toxic, industry has often proceeded to find replacement materials, sometimes with a strong push from safety activists. The 2016 Frank R. Lautenberg Chemical Safety for the 21st Century Act calls for greater public awareness of chemical information and risk-based chemical assessments, among other important measures, to improve the nation's chemical management.[5]

Today, many consumer product manufacturers have stopped using plastics containing BPA altogether (not just in baby bottles) and conspicuously tout them as BPA-free, presumably to reassure consumers and compete for business on the basis of being "green." Some consumers have even switched from using plastic containers of any type to using food-grade stainless steel containers.

The gases and vapors that can be emitted from spacecraft materials concerned NASA enough for it to develop a standard on material outgassing properties, with the goal of creating a healthier environment for astronauts.[6] *(Photo: Colourbox.com)*

Choosing materials

With so many molecules and compounds going into various types of consumer and industrial products, designers face quite the challenge to understand the risks. The best bet is for designers to remain aware of materials suspected of ill health effects and select those considered safer. While chemical leaching is a concern, so is outgassing. What kind of materials are prone to outgassing? The list is lengthy and includes plastics, adhesives, insulation, and cleaners. Logically, designers should use materials that have less of an outgassing problem or do not outgas at all.

"Hazardous substances, such as some plastic monomers, solvents, additives and byproducts...can be released during all phases of plastic life cycle..."[7] In fact, a study assessing the harmfulness of plastic polymers found that 29% of the 55 studied polymers contained monomers classified as either mutagenic, carcinogenic or toxic for reproduction.[7]

Unfortunately, some materials in common use can create a toxic air around them...

That "new car smell" might be pleasing to some people, but like the gases produced by some air fresheners, the chemicals within are not necessarily good for one's health.[8] Reportedly, new automobile interiors are constructed of materials that collectively can contain hundreds of different chemicals, some of which outgas and have not been safety tested.[9] The source of that unique scent "is in the various solvents, adhesives, plastics, rubbers and fabrics used in car construction. Many of these contain volatile organic compounds (VOCs), some of which can be deadly in sufficient quantities. Others are just bad for you."[10]

FREE
with every vechcle!
THAT NEW CAR SMELL!!

Exemplar 1
Car seat

Car seats have the hefty responsibility of protecting children from injury or death during a car crash. They incorporate some self-evident safety features, as well as some less obvious (but important) ones.

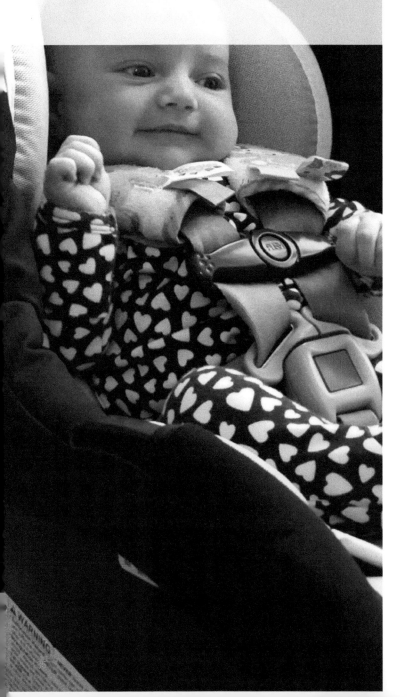

Include pads /
Make things easy to clean
Principle 5 - pg. 23 / Principle 2 - pg. 17

Padding is reinforced with foam that provides multi-directional protection during collisions. The seat fabric and padding can also be removed for cleaning without having to disassemble the harness.

Provide restraints
Principle 95 - pg. 221

Five-point harness keeps the child secure in the case of an accident. Adjusting the harness does not require disconnection, thereby reducing potential for incorrect assembly.

Make it fire resistant
Principle 9 - pg. 31

Fire-resistant fabric lining and foam padding reduces potential for deaths and injuries associated with vehicle fires (and meets federal flammability requirements).

Encourage safe lifting / Provide stabilization

Principle 34 - pg. 87 /
Principle 1 - pg. 15

Handle located above the car seat's center of gravity provides a stable, balanced grip. The low center of gravity over the product's "footprint" prevents the car seat from tipping over when set down.

Add conspicuous warning

Principle 79 - pg. 185

Conspicuous warning labels remind users to avoid placing a rear-facing car seat in the front passenger seat, to use the product for children only within a specified height and weight range, and to utilize the restraints properly.

Provide tactile feedback

Principle 8 - pg. 29

Handle "snaps" into several distinct positions, emitting a tactile click to indicate that the handle is locked into place. The car seat also "clicks" firmly into the base to indicate that the seat and base are securely attached.

Lock touchscreens

Principle
Enable users to lock touchscreen-based user interfaces to prevent unauthorized access and/or accidental actions.

Lock it up

In most cases, if a touchscreen conveys safety-critical information and functionality, it should have a screen lock. The feature will protect the screen from inadvertent touches that could lead to undesirable outcomes. It might also prevent intentional interactions by unauthorized individuals. Additionally, a screen lock enables users to safely clean the display without actuating on-screen controls.

If the only concern is preventing unintended actuations, the screen lock could take the simple form of a virtual button that must be slid across by means of a gesture, similar to the Apple iPhone's patented[1] *Slide to Unlock* feature.[2] Like Apple's feature, the screen could also provide instructions on how to unlock it using the gesture, thereby reducing the potential safety consequences resulting from someone not being able to unlock it. To prevent unauthorized access, user interfaces can require users to enter a passcode to unlock the screen.

Note that critical control actions might also warrant additional protections, such as confirmations and possibly even automated fail-safe responses.

A factory worker could bump a touchscreen and command material-handling robots to move unexpectedly and/or in a dangerous manner.

New types of authentication

To simultaneously make touchscreen locks both more secure and faster to unlock, tech companies continue developing new types of user authentication methods like fingerprint sensors and face recognition algorithms.

In 2017, Apple introduced the iPhone X, which features facial recognition technology that can reportedly recognize the user's face and unlock the phone in just 3 milliseconds. The iPhone X camera projects and analyzes more than 30,000 invisible dots to create a precise depth map of your face and is designed to protect against spoofing by photos and masks.[3]

Automatic lockouts

Have you ever forgotten to lock your house or car door before? Of course—we all have. Rather than rely on users to remember to manually lock a touchscreen, many touchscreens lock out after a preset duration. For example, many cell phones lock out after 1 minute of inactivity to prevent unauthorized users from accessing the phone's contents.

There are some cases in which it would be safer to automatically lock the screen during certain situations and not just after a period of inactivity. For example, some touchscreen-based car infotainment systems automatically lock while the car is in motion to prevent the driver from interacting with, and being distracted by, the touchscreen.

 In 2015, the number of deadly crashes involving distracted driving increased 8.8% to 3,477. Research suggests the prevalence of touchscreens in cars might be a contributing factor.[4]

Touchscreen locks in healthcare

When you hear the word "touchscreen," you might first think of smartphones and computer tablets. But, touchscreens are now ubiquitous in many industries. While a screen lock on a phone helps protect an individual's privacy, screen locks on specialized devices can help prevent other harmful situations. For example, imagine a scenario in which a patient monitor keeps alarming. Without a lock feature, a well-meaning family member might adjust the alarm limits so the monitor doesn't disturb his/her sleeping loved one, thereby creating a harmful situation where the monitor might not detect a dangerous change in the patient's condition. To prevent this exact situation (and others in which someone might not have good intentions), some critical care devices that are likely to be in patient rooms have password-protected screen lock features.

Now consider a touchscreen-based defibrillator being used in a rescue scenario. A first responder carrying the device while rushing to the scene or trying to use the device in a jostling ambulance could accidentally bump the touchscreen or press the wrong button. A potential mitigation is to automatically lock the touchscreen when an accelerometer detects that someone is carrying the device. Another potential mitigation is a physical unlock button that requires users to take deliberate action to unlock the touchscreen.

Make buttons large

Principle

Make buttons large and ensure there is sufficient space between buttons to facilitate efficient use and prevent unintended button presses.

Make buttons large...

Pressing the wrong button—everyone has done it, and usually without consequences greater than causing an elevator to stop at the wrong floor or switching to the wrong TV channel. But, sometimes the wrong button press can be more problematic and dangerous, such as turning on the front right—rather than front left—burner on a stovetop range, potentially causing a burn or fire. Similarly, a wrong button press might deactivate an automobile's traction control system instead of turning on the rear window defroster.

Although making a button large cannot prevent mistakes wherein someone deliberately presses the wrong button, it can prevent unintended button presses due to poor aim and/or close proximity between buttons. A large button doesn't require accurate aim and can even accommodate people who might not give the "pressing" task their full attention. Also, a large button naturally increases the center-to-center spacing between adjacent buttons, making it less likely that someone with a broad fingertip (i.e., finger pad) unintentionally presses a neighboring button.

...but not too large

Keep in mind that making buttons too large can also come with its fair share of consequences. Large buttons that protrude from a system or control panel are more susceptible to inadvertent actuation due to accidental bumps. This is particularly true in fast-paced environments where operators perform actions quickly, or in environments with high amounts of movement throughout (e.g., a crowded ICU in which clinicians transport or reposition portable hospital equipment within a small space).

Designing the perfect pushbutton

When designing a pushbutton, consider the following guidelines to help ensure rapid and accurate button presses:[1,2]

- **Size.** Design buttons to be at least 0.8" wide.

- **Force.** Keep button resistance below 8.9 N (2 pounds) to ensure a majority of users can press it successfully.

- **Feedback.** Ensure the button provides immediate auditory and/or tactile feedback when pressed.

- **Travel.** Enable the button to depress (i.e., travel) about 0.12" when pushed to provide adequate tactile feedback.

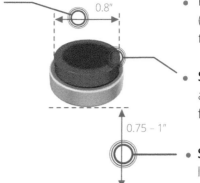

- **Labeling.** Include a meaningful label (with text and/or icons) on or above the button to indicate its function.

- **Surface.** Use a concave surface with a small lip on the edges to prevent fingers from sliding off.

- **Spacing.** Space adjacent buttons at least 0.75 – 1" apart.

- **Backlighting.** Provide a backlight to indicate when a button is active and/or to make it visible in dim lighting.

Many of the guidelines described above also apply to software; larger, on-screen buttons with sufficient spacing will likely increase the accuracy of button presses, particularly on small smartphone screens. Note, however, that hardware button guidelines might vary depending on the button type. For example, keyboard buttons don't need to be spaced as far apart from one another because they are often used in a complex, sequential fashion.[1]

Fitts's Law

Fitts's Law is a model related to "pointing" movement (i.e., accessing a target with a finger or a computer pointer) that describes the relationship between (1) the distance to the target, (2) the size of the target, and (3) the amount of time it takes to reach the target. The model states that the shorter the distance to the target and the larger the target itself, the faster a user's hand or pointer can reach the target accurately.[3]

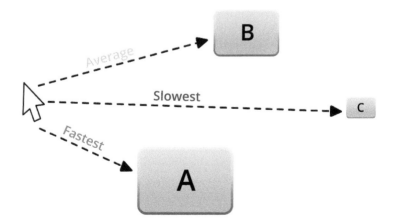

As you might expect, incorporating Fitts's Law into button design for safety-critical tasks can enable efficient use and yield potential safety benefits. For example, an "emergency stop" button should be sufficiently large and relatively close to the work area (although not vulnerably placed) to enable a user to access it quickly and accurately in the event of an emergency.

Use consistent units of measure

Principle

Use appropriate and consistent units of measure to reduce the chance of confusion (e.g., mistaking a value measured in kilograms as one measured in pounds) and, consequently, taking the wrong action.

Mix-up mishaps

The parallel existence of the US customary system, the metric system, and other measurement systems has been the cause of tragic and costly mishaps.

In September 1999, NASA's Mars Climate Orbiter was supposed to enter orbit around the red planet with grace. Instead, the $125 million craft's approach was too close to the planet, leading it to plunge through the planet's atmosphere and disintegrate. The mishap's cause turned out to be a mismatch between the units of measure (US customary versus metric) used by two companies collaborating on the mission.[1] The mismatch placed the orbiter tens of kilometers closer to the planet than intended.[2]

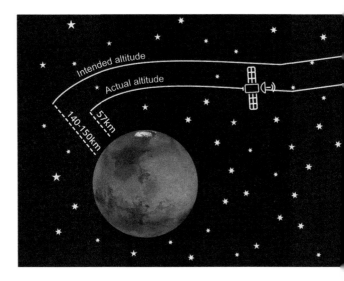

In July 1983, Air Canada Flight 143 transformed itself from a jet-powered aircraft into a glider after the crew performed a calculation confusing pounds and kilograms. The mistake led them to think they had 20,400 kg of fuel, when they actually had only 9,144 kg.[3] The aircraft, a Boeing 767, ran out of fuel at 41,000 ft—a high enough altitude to enable a 60-mile glide that culminated in a rough landing at a closed air force base in Gimli, Manitoba. Hence, the aircraft earned the nickname "Gimli Glider." Fuel gauge malfunctions eliminated a possible means of detecting the error before it created an emergency.[4]

Preventing mishaps

As seen by these mishaps, it is particularly important to use appropriate and consistent units of measure, ideally sticking to one of the two common systems and then avoiding mixing different measurements within a single measurement system (e.g., feet and yards). When using two measurement systems is unavoidable, the units of measure should be clearly indicated in close proximity to the measurements. In some cases, accidents can be avoided by having a computer perform automatic checks to determine if data is appropriate (i.e., in the expected safe range) and, if it is not, require the user to confirm a value.

For instance, an electronic health record (EHR) used to document a newborn child's physical traits could compare the entered body length of 20 in. with the entered weight of 3.4 lbs. Then, the EHR could determine that the weight was probably entered in kilograms rather than pounds because the expected weight for a child of that length should be closer to 7 or 8 lbs—much higher than the entered value of 3.4.[5]

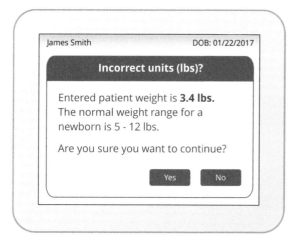

In some cases, computers can perform automatic checks to see if data entered falls within the appropriate range depending on the "norm," and then ask the user to confirm values outside the norm.

Another common mix-up involves calculating a medication dose based on a patient weight specified in the wrong units of measure. In one case, a toddler's weight was incorrectly entered into an electronic medical record as 25 kg, when it was actually 25 lbs—an overstatement of 220%.[6] In one study, this type of error, or the reverse case in which kilograms were used in a calculation instead of pounds, accounted for 26.9% of medication errors based on patient weight.[7]

Remove humans from the equation

Making people convert units is a recipe for disaster. People can be fairly unreliable at conversion tasks, whether they do the math in their heads or with a calculator. People can transpose digits, use incorrect conversion rates, and make erroneous key presses. That is why it can be helpful for products, such as digital scales, to enable users to shift or convert between units of measure via a single button press.

Ambiguity over units of measure, as well as accommodation of individuals familiar with one system of measure versus another, can also be resolved by computer systems that can automatically convert the values and display multiple units at once. The best solution, however (as described throughout this chapter), is to use a single system of measure in a given location or workplace.

Some computer systems can automatically convert the values in real time and display multiple units at once.

Ensure strong label-control associations

Principle

Make sure user interface controls are clearly grouped with their corresponding labels.

Is that label for this button, or...

Labels help users identify physical items such as displays, controls, and mechanisms. They also help users identify information that is presented on digital displays. In fact, one could argue that all physical items and on-screen content, perhaps except for items that serve intuitively obvious purposes, should be labeled to ensure correct identification and interpretation.

Unfortunately, poor labeling and especially poor label placement can trigger harmful mistakes. Users can commit use errors such as basing decisions on the wrong information and activating or deactivating the wrong function. Imagine a respiratory ventilator that has pediatric and adult modes. Choosing the adult mode with a deliberate press of the wrong button, due to confusion about which among two buttons is used to select the pediatric mode, could cause fatal injury to a child.

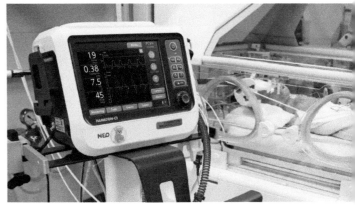

Respiratory ventilators often have adult and pediatric modes that should be clearly labeled.

Florida's butterfly ballot from the 2000 US Presidential election.

Poor label-control association in voting ballots can cause people to accidentally vote for unintended candidates. The candidate labels in Florida's butterfly ballot during the 2000 US Presidential election did not clearly correspond with the punch holes in the middle, which confused many voters.[1] These voters thought they voted for Al Gore, whose name appeared second in the list of candidates, but whose punch hole was third in its column, thereby resulting in a vote for Pat Buchanan.[2]

Label placement guidelines

- **Place the label on the control.**
  Combining the control and label into a single element creates a direct association between them and simplifies the user interface's appearance.

- **If there isn't enough space on the control for its label, place the label above the control.**
  Placing a label below the associated control almost ensures that the label will be blocked by the user's hand when the user presses the control. *Note: There might be a compelling reason to place a label below a control when its position below the control helps communicate a fore-and-aft or up-down physical relationship of the associated components or functions.*

- **If placing a label above a control is not practical, place it to the left rather than to the right.**
  When designing for users who read left-to-right, placing the label to the left can create a clearer association with the control. This also helps to prevent right-handed individuals (about 90% of the population)[3] from obscuring the label with their hand.

- **Place the label closer to the associated control.**
  Placing a label equidistant to two controls can lead to selection errors, particularly when the user is working quickly. To create clearer associations between a control and its label, make sure the label is placed close to the control, and that controls are spaced far enough apart to be seen as separate.

- **Create functional grouping.**
  A user interface with a large number of controls can be difficult to process and can appear daunting. Labels can be visually grouped together to separate a functionally-related group of controls from other controls. Using a hierarchical labeling scheme, whereby specific component labels are complemented by an overall label, can also be advantageous. For example, a group of controls might be labeled "Pump A," and the individual controls might carry the labels "RPM," "Oil", and "Flow Rate."

Worse   *Better*

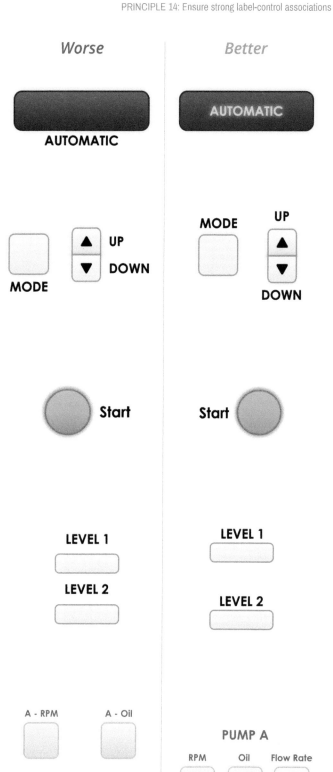

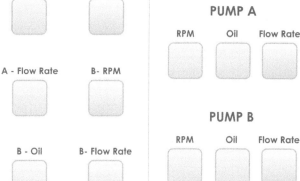

Provide reminders

Principle

Provide clear, conspicuous, well-timed, and specific reminders to prevent users from forgetting to perform important actions.

Forget me not!

We forget things all the time. While forgetting to turn off a light when leaving the house usually results in annoyance and nothing more, forgetting to deploy airplane wing flaps before takeoff can cause a crash. Safety can depend on people doing the right thing at the right time. A reminder helps ensure this happens.

This is certainly true for a skydiver in freefall. It's unlikely that the skydiver will forget to pull the rip cord to deploy his or her chute. But, it is possible for him or her to get distracted and not leave enough time to deploy the main chute. That's where an Automatic Activation Device (AAD) comes into play. It's designed to deploy a reserve chute if the skydiver descends below a specified altitude. Similarly, many skydivers use altimeters that alarm below a minimum altitude[1] (see *Principle 28 – Add a horn, whistle, beep, or siren*).

Checklists are another type of reminder, usually to perform several tasks (including inspections), and possibly perform them in a prescribed sequence. They are so effective—as long as the user does not forget to refer to them at the proper time—that airplane pilots are expected to follow them every time they take off. The need for a preflight checklist stems from the fact that pilots need to do many things before taking off, including several tasks that, if skipped, could lead to disaster. As stated before, many aviation accidents have been caused by a pilot's failure to partially deploy the wing flaps to generate more lift during takeoff. Lacking the necessary lift, planes without extended flaps have failed to take off, or have stalled once barely in the air.[2]

Checklists are also used before surgical cases to ensure that clinicians confirm the patient's identity, which body part is being operated on, and if the patient has any allergies that need to be considered when administering drugs. "Timeouts" during which clinicians follow their checklists have helped prevent mistakes such as the wrong limb being amputated.[3]

A healthy reminder

Many healthcare devices provide reminders to help users manage their treatment and maintain their devices. For example, continuous blood glucose monitoring (CGM) systems provide people with diabetes with real-time glucose readings and glucose trends that help them better manage their diabetes. But, if CGMs are not calibrated regularly, the systems can produce inaccurate readings and frequent false alarms. So, such systems typically provide calibration reminders approximately every 12 hours, reminding users to measure their blood glucose with a glucose meter and use the result to calibrate the system.[4]

Too much of a good thing

Have you ever swiped away a notification on your phone, or closed a pop-up window without even reading its contents? This desensitization to reminders— a close cousin of "alarm fatigue"— happens to everyone, but when it happens to doctors and nurses, the result can be more dangerous than a missed appointment.

Most hospital electronic health record systems display so many safety notifications (i.e., alerts) that clinicians tend to ignore them. In fact, studies suggest that clinicians ignore safety notifications between 49% and 96% of the time.[5] In one instance at Children's Hospital of Philadelphia, doctors ignored a relevant reminder about a patient's drug sensitivity. As a result, the amount of medication the patient received induced a potentially lethal reaction.[5]

Model photo: Colourbox.com

Design reminders to be...

Conspicuous
Make reminders visually and/or audibly distinct. Account for environmental noise and distractions to make sure users notice reminders.

Well-timed
There's no point in providing reminders after they're pertinent or actionable, but also no point in providing them too soon—before users can take preventative action. Time reminders wisely.

Clear
Use clear and succinct language to ensure that users can interpret and respond to reminders quickly and correctly.

Actionable
Clearly present the action(s) users can perform when responding to the reminder and guide the user to complete the action(s) correctly.

Smart
Prevent reminder fatigue and habituation by filtering out reminders for minor/non-critical tasks. Only provide reminders for critical tasks.

Make decimal values distinct

It's a giant decimal point... get it?

Principle
When designing numeric displays, make the decimal points large and remove "trailing zeros" to reduce the likelihood of a user overlooking a decimal point or misreading a critical numerical value.

Small decimal points are a big deal
The digital landscape is chock-full of numbers with decimal places. This presents an increased opportunity to misread numbers by overlooking decimal points. One cause of misreading numbers can be the small size of the decimal points. In many cases, decimal points are the size of a single pixel or just a few pixels, or a dedicated segment on a segmented display that is quite small. In the case of dot matrix displays, the decimal point might be the same size as the dot over the letters "i" and "j." In these cases, the designers have missed the point (pun intended) that a decimal point is an important character that should be made larger for legibility's sake.

When are large decimal points necessary?
Cases that merit an oversized decimal point are those in which a user will base important actions on a displayed value, such as a vital sign presented on a patient monitor, the rate when programming drug infusion pumps, or RPM (revolutions per minute) readings on power plant control panels and aircraft cockpit displays. For example, a nurse who overlooks the decimal point on a patient monitor might read the $EtCO_2$ value as 35 mmHg (a normal rate)[1] rather than 3.5 mmHg and not realize that a patient's endotracheal tube is obstructed, resulting in respiratory distress.

Worse   ### Better

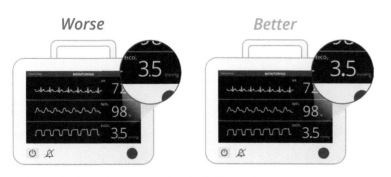

Enlarging the decimal point can reduce the likelihood of a nurse overlooking the decimal point on a patient monitor and misreading the $EtCO_2$ value as 35 mmHg (a normal rate) rather than 3.5 mmHg.

Decimal points in dot matrix displays
Some users might also be more likely to recognize a decimal point if it is round, rather than square. Displaying a round decimal point on a segmented display or LCD display is relatively straightforward (just display a circle). It's slightly more complicated on a dot matrix display. On a dot matrix display, a round dot can be approximated by starting with a square and then adding rows to each side of the square that are progressively shorter. The shorter rows create a rounding effect.

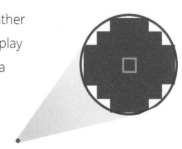

Techniques for preventing numerical misreadings

The challenge for designers and programmers is to display whole and decimal values such that they are visually distinct and not vulnerable to being misread. Finding a solution, such as creating greater separation between numbers, might be an essential step to prevent misreadings. Another way to help ensure that numbers with decimals and following digits are read correctly is to make the digits after the decimal point smaller, thereby differentiating whole values from decimal values. Additionally, designers should "decimal-align" a vertical list of numbers (i.e., line up the decimal point in each value), rather than left-align the first digit, to help users differentiate between whole and decimal values.

Worse

Infusion Prescription

CAROlyne 250mg

Flow rate: 2.5 mL/hr
Volume: .75 L

Better

Infusion Prescription

CAROlyne 250mg

Flow rate: 2.5 mL/hr
Volume: 0.75 L

Enlarging the decimal point and reducing the size of the digits following the decimal point can reduce the likelihood of a user mistaking whole and decimal values, such as those on a patient's infusion prescription. In this case, for example, misreading the flow rate as 25 mL/hr rather than 2.5 mL/hr can result in an incorrect flow rate and potential harm to the patient.

Note: Many countries in Europe and other parts of the world use a comma instead of a decimal point, suggesting that the commas in these cases also need to be sufficiently large to ensure that they are noticed.

Trailing zeros

In addition to the techniques above, designers should take caution when using a "trailing zero" (i.e., a zero after which no other digits follow) when writing numerical values. The numbers "300.0" and "30.0" have trailing zeros. So, what's the problem?

The problem is that people can overlook the decimal point and misread the numbers above as "3000" and "300," respectively. One solution is to remove the decimal point and trailing zero altogether if the value can be expressed as a whole number—that is, if the risk of misreading the value outweighs the benefit of the added precision of the trailing numeral. Removing the trailing zero might work for larger whole numbers, but it might not be possible for ones as small as "3.0" because the added precision (e.g., the difference between "3.3" and "3.7") might be essential. This is often the case with hospital lab values where the higher precision is needed to assess a patient's condition accurately.

Zeroing in on prescription errors

Trailing zeros can be particularly hazardous when it comes to drug prescriptions. For example, writing a medication prescription as "1.0 mg" can be misinterpreted as 10 mg, thereby leading to a 10-fold medication overdose.

However, it's not only trailing zeros that can pose an issue. A lack of leading zeros can also lead to medication errors. For example, a prescription written as ".1 mg" can be misinterpreted as "1 mg" if a user overlooks the decimal point—again, leading to a 10-fold medication overdose. In this case, adding a zero before the decimal point and presenting the dose as "0.1 mg" might make the dose less prone to misinterpretation.

Display real-time data or use time stamps

Principle

Continuously update or time-stamp critical data so users do not make critical decisions based on old or inaccurate data.

Live data

Many critical decisions and actions are driven by data presented on product displays. Such data includes blood test results for a critically-ill patient, the pressure in a power plant boiler, or an airplane's airspeed.

Updating data in real time (or at least as frequently as needed to support a given user's task) is probably the best approach when it comes to presenting safety-critical data.[1] Such "live" data enables people to make decisions and react to a situation as it unfolds without lag (i.e., latency). "Stale" data or latency might interfere with effective monitoring and delay proper action (or even induce people to take the wrong action if they are reacting to inaccurate data). Clearly, an aircraft's airspeed should be updated continuously to give pilots the information they need to fly safely, as opposed to overspeeding or stalling in various stages of flight.

A car's dashboard displays the car's speed in real-time, which enables the driver to adjust the car's speed as needed.

Static data

Depending on the use scenario and the frequency of data changes, a short delay might or might not be acceptable. Less frequent updating of slow-changing data might be advantageous because it cuts out distracting and relatively meaningless changes compared to an averaged value that gets updated every 10 minutes, for example. In other cases, less frequent updating might be needed to conserve battery life. However, when updates are less frequent, critical data should be time-stamped so that the user knows its age and can determine when "fresh" data will become available. Moreover, in some cases, the user should even be able to refresh the data on-demand, rather than wait for the data to refresh automatically after a pre-set duration.

Time-stamp your data!

What is an appropriate elapsed time threshold for including or excluding a time stamp? There is no specific threshold because the right one will depend on the type of data and how it is being used. In the case of a device programmed to take non-invasive blood pressure (NIBP) measurements, it could take 40 seconds to inflate a pressure cuff and measure the systolic and diastolic pressure points, so continuous updating might not even be possible.[2] One vital signs monitor allows users to set the NIBP measurement interval over the following range: 1, 3, 4, 5, 10, 15, 30, 45, 60, 90, 120, and 240 minutes.[3] Knowing if pressure readings are 1 or 240 minutes old could be safety-critical considering that a healthcare provider might base blood pressure medication on the value. Of course, the healthcare provider could always perform a new "STAT" (i.e., on-demand) measurement to ensure he/she medicates the patient properly.

One automatic blood glucose measurement system used at a patient's bedside works similarly to the NIBP monitor described above—it measures the patient's blood glucose value at regular intervals. The system displays the last measurement time right next to the current time of day, enabling the healthcare professional to compare the time stamp with the current time to determine the measurement's age.

Similarly, hand-held glucose meters used by people with diabetes display blood glucose measurements next to the time and date on which they were recorded. This helps prevent the mistake of delivering too little or too much insulin based on an outdated and no-longer-accurate blood glucose measurement.

A blood glucose meter displays the time and date the measurement was taken.

Creating an effective time stamp:

- **Clearly distinguish the time stamp from the current date/time.**

- **Include "AM/PM" if you use a 12-hour clock format.**

Group the time stamp and the associated data; in this case the time stamp is grouped together with the fetal ultrasound data.

Clearly distinguish day and month if presenting a date.

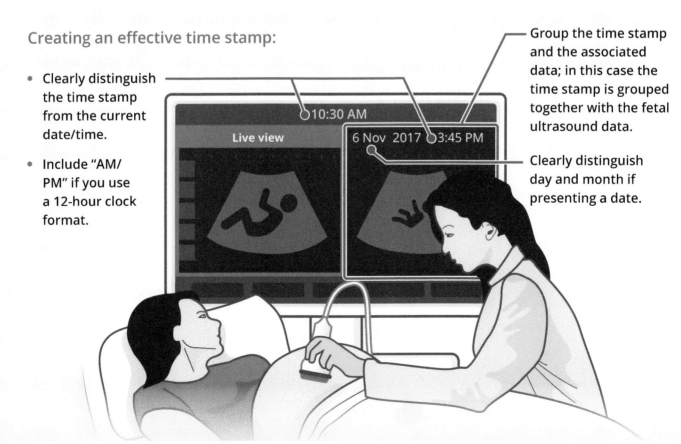

Predict hazardous situations

Principle

Products should give forewarning so users have a chance to prevent or avoid hazardous situations before they can occur.

What is a predictive system?

Predictive systems can determine where things are heading and can give fair warning, and, in some cases, automatically change course to avoid danger. Logically, the value of a predictive system depends on its accuracy. Avoiding false positives (i.e., predicting a hazardous situation that will not occur) and, perhaps more importantly, avoiding false negatives (i.e., not predicting a hazardous situation that unfolds) is key.

Predicting hazards in our bodies

Continuous blood pressure monitors can detect when a patient's blood pressure is rising toward a dangerously high value and issue an alert to take action.

Hospital beds can alarm when a patient moves or attempts to exit the bed so that a nurse can come to assist and protect against a fall.

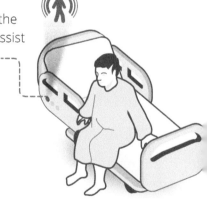

Insulin delivery systems that project a user's near-future blood glucose level offer users the option to adjust dosing to avoid dangerously high or low blood glucose levels.[1]

Predicting hazards in our environments

Software algorithms fueled by the right data can be used to help identify buildings at risk of fire, giving city inspectors a chance to inspect potentially dangerous properties and intervene as necessary.[2]

Airline warning systems issue a ground proximity warning, "Woop Woop – Terrain Terrain," if the current flight path will lead to a crash.

Automobile collision avoidance systems can detect hazards (e.g., an object in the road) and automatically apply the brakes to avoid a collision.

Enabling users to respond effectively to predictions

When designing a predictive system, consider the following stages:

1. Detect
Make sure the product detects signals quickly and accurately, while avoiding false signals that might result from short-lived, insignificant changes.

2. Notify
Provide conspicuous notifications, possibly using multiple channels (e.g., visual, audible, and vibratory signals) to get the user's attention.

3. Inform
Provide a concise and clear explanation of the detected signals and the associated prediction.

4. Advise
Clearly present the different actions the user can take in response to the prediction, and empower the user to take corrective action.

5. Guide
Guide the user to perform the suggested action correctly.

6. Update
Update the user regarding the impact of his/her actions—did the user avoid the hazardous situation, or does the user need to take additional action?

Prediction: we won't need to do anything

Even with the aid of predictive systems, humans might be too slow or unreliable (on average) to avoid certain hazards effectively, leading some designers to consider automation as a means to compensate for our shortcomings. For example, some automobile manufacturers are rapidly trying to develop fully-automated (i.e., self-driving) vehicles rather than deal with the "almost insurmountable engineering, design, and safety challenges" of passing control in an emergency situation to a potentially distracted (or even sleeping) human.[3] Full automation might not be possible or appropriate in certain contexts, but it presents a promising new way of eliminating risk through design and technology.

Make software secure

Principle
Implement cybersecurity measures in digital or network-based systems to protect critical software from being hacked.

The problem with hacking

In the late 1990s, there was some panic and a lot of remedial work performed to avoid harms related to the Y2K (i.e., Year 2000) problem—that is, computer code written in the 20th century, which used two-digit year indicators, not being able to recognize that "00" referred to "2000" rather than "1900." Ultimately, there were no widespread problems or stories of Y2K-related tragedies.

A more insidious problem is critical software being hacked in ways that could cause great harm. Consider the havoc that bad actors could wreak if they hacked into power generation, transportation, or hospital systems, or even people's homes (e.g., Internet of Things (IoT) devices), to name just a few targets. In the case of hospitals, a bad actor could theoretically change the programming of life support equipment or pose a threat to lives and demand a ransom.

The government's role

Regulators and government agencies have been and remain concerned about the vulnerability of digital systems and critical medical devices. For example, the National Institute of Standards and Technology (NIST) released the Digital Identity Guidelines in June 2017, which includes guidelines for ensuring the "proofing and authentication of users... interacting with government IT systems over open networks."[1] And, on July 31, 2015, the US Food and Drug Administration (FDA) issued a warning about using a particular manufacturer's infusion pump due to the threat of it being hacked. An online safety communication to the public read, "The FDA is alerting users of the [brand of infusion pump] to cybersecurity vulnerabilities with this infusion pump. We strongly encourage that health care facilities transition to alternative infusion systems, and discontinue use of these pumps."[2]

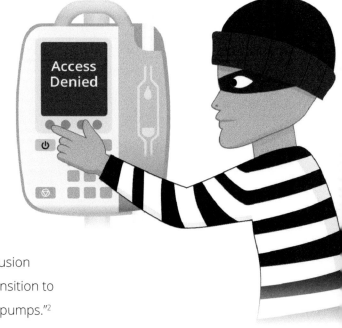

"Sorry, 'password1234' is not an acceptable entry"

Threats to software security might be local (e.g., disgruntled employees) or external (e.g., lone actors, nation states). Managing risk is complex, involving cybersecurity measures such as encryption, control over access (e.g., authentication), anti-virus software, software updates, intrusion detection, and the ability to evolve to match dynamic threats. In an increasingly connected world filled with IoT, products must include defenses against their corruption. Some protective measures are user-facing and increase a product's security by affecting users' interactions with the product at hand. Sample design guidelines for these front-end solutions include:

✔ Requiring users to enter a password to view sensitive information, such as patient details or financial information.

✔ Instructing users to set a password that has a high "password strength" (i.e., is less likely to be hacked), such as one that features more than 8 characters and a mix of uppercase and lowercase letters, numerals, and symbols.

✔ Preventing users from setting a password that does not meet basic security guidelines (e.g., using words found in a dictionary).

✔ Instructing or requiring users to change their passwords on a regular basis.

✔ Automatically logging out a user after a certain amount of idle time.

✔ In physical products, requiring the use of a special key, interlock, or even biometric sensors to ensure that the user is authorized to operate the device (see *Principle 26 – Incorporate a lockout mechanism*).

FreQu3nTly U$ed P@$sword$

According to a public data source of 10 million passwords, the top 10 most common passwords from 2016 are:[3]

1. 123456
2. 123456789
3. qwerty
4. 12345678
5. 111111
6. 1234567890
7. 1234567
8. password
9. 123123
10. 987654321

Make sure your password isn't on the list!

The great debate: security versus usability

As you read the items in the list above, you might have been thinking about how frustrating it can be to create a brand new password each time a website or application requires different password security requirements. Or, you might have thought about how difficult it is to remember all of the different passwords you create. But, security and usability do not necessarily need to be mutually exclusive. Manufacturers can achieve high security to promote safe use of their systems and create a satisfying user experience simultaneously when they consider security features and design from the beginning of development, rather than adding password requirements as the last step, for example. In some cases, early brainstorming can lead to hardware developments that increase usability and nearly eliminate the need for password entry altogether, such as the Apple iPhone's Touch ID (built-in fingerprint reader)[4] and Face ID (built-in facial recognition) technology.

Do not require mental calculation

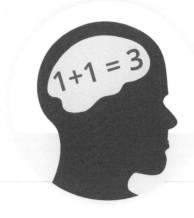

Principle

Products should perform calculations automatically, rather than require users to make mental calculations that could involve errors and lead to harm.

The danger of miscalculations

Regardless of whether a particular individual is "good at math," he or she can be prone to make a calculation error in his or her head at some point, and the error could have significant consequences. Mental math is not too difficult when multiplying 2 x 100 to get 200. It is more challenging when multiplying 43 x 7 to get 291 or dividing 700 by 4 to get 225.

To prove a point...the answers to these calculations are actually 301 and 175, respectively.

Sometimes, getting the math wrong causes inefficiencies, annoyance, and embarrassment. Other times, it could spell trouble. An example of the latter is performing mental calculations when load-balancing a cargo vehicle. The wrong balance could lead to dangerous vehicle dynamics.

Another example of a mental math error leading to trouble is someone trying to determine how much medication (e.g., insulin) to take based on a blood glucose measurement. Get that math wrong and the result could be severe hypoglycemia or diabetic shock, which can lead to a coma.

Some people have a condition called dyscalculia—a mathematical disorder that makes it difficult to understand numbers and numerical concepts (e.g., telling time), and to perform calculations. Researchers estimate that 3 to 6.5% of children in the general school population have the disorder,[1] which can extend into adulthood. A genetic form of the condition has been linked to functional and structural anomalies of the brain's parietal lobe.[2]

Provide calculation tools

When the consequences of a math error are safety-critical, products should not require users to perform mental calculations or even calculations supported by pencil and paper. Instead, microprocessor-endowed products should provide calculation tools—perhaps a step-by-step wizard that directs the user to enter specific values and then performs the necessary calculations, providing guidance in a user-friendly manner (e.g., "Deliver 2.5 units of insulin"). It might be safer still for the product to check the value against norms to ensure that it is in a normal, expected range. And better yet, a product should not require users to perform calculations when the operation can be automated, or if necessary values can be presented in their final form and not require any math.

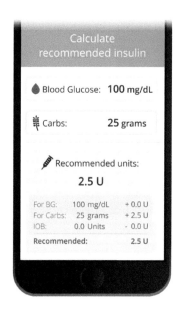

Here is a case of a product eliminating the need for mental calculation: a product might display that its power reserve is down to 35%. The user might be aware that a full charge lasts 5 hours. To figure out when the device will run out of power at an average drain rate, the user must calculate 35% of 5 hours (e.g., 0.35 x 5 = 1.75 hours). But, it would likely be much more reliable and easier for the user if the product displayed: "Estimated power remaining: 1 hour, 45 minutes."

Real-world example: Smart Infusion Pump

Smart infusion pumps can automatically calculate infusion values that clinicians have traditionally calculated manually. For example, such pumps can calculate the dose if a user enters an infusion rate. The pump can then check the calculated dose against safe limits and notify the user if the dose is outside safe limits.

Worse

Infusion pump allows the user to start an infusion with a very high dose, which could be dangerous for the patient.

Better

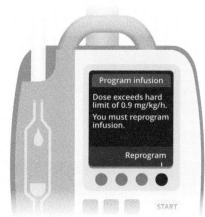

Infusion pump checks dose and notifies the user that the dose is slightly above limit, but allows the user to proceed.

Infusion pump checks dose and notifies the user that the dose is above the hard limit and does <u>not</u> allow the user to proceed.

Exemplar 2
Diabetes management software

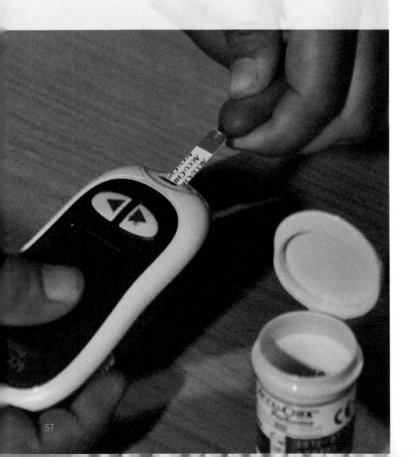

Many people with diabetes manage their condition by using software applications that include features such as food-tracking, blood glucose value-tracking, and an insulin dose calculator. Such applications, which might reside on a computer or mobile phone, are transforming self-care while also enabling clinicians to monitor their patients more closely.

Provide reminders / Predict hazardous situations
Principle 15 - pg. 45 / Principle 18 - pg. 51

Alerts warn users when they log a particularly high or low blood glucose value—one that could lead to a medical emergency (e.g., unconsciousness). Such alerts suggest what steps users should take to take to avoid harm.

Use consistent units of measure
Principle 13 - pg. 41

Constraining software to match a glucose measurement device's units of measure (e.g., using only mg/dL, rather than both mg/dL and mmol/L) prevents mix-ups that could lead to incorrect insulin dosing.

Make buttons large / Ensure strong label-control associations
Principle 12 - pg. 39 / Principle 14 - pg. 43

Large, well-spaced, and clearly labeled buttons help prevent users from accidentally pressing the wrong one.

Make software secure
Principle 19 - pg. 53

Password-protected software helps prevent bad actors from gaining access to private, health-related data, and possibly even altering critical settings in a manner that could cause harm.

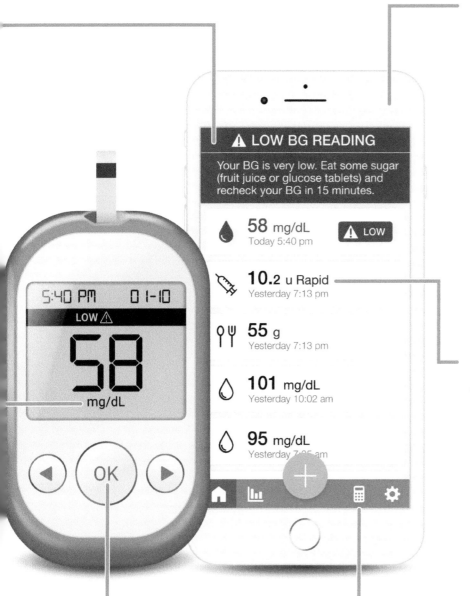

⚠ LOW BG READING

Your BG is very low. Eat some sugar (fruit juice or glucose tablets) and recheck your BG in 15 minutes.

58 mg/dL
Today 5:40 pm   ⚠ LOW

10.2 u Rapid
Yesterday 7:13 pm

55 g
Yesterday 7:13 pm

101 mg/dL
Yesterday 10:02 am

95 mg/dL
Yesterday 7:05 am

5:40 PM 01-10
LOW ⚠
58
mg/dL

OK

Make decimal values distinct
Principle 16 - pg. 47

Distinct decimal values can prevent users from misreading critical values. For example, misreading 2.5 insulin units as 25 insulin units could potentially be a deadly mistake.

Do not require mental calculation
Principle 20 - pg. 55

Bolus calculators help the user determine the correct amount of insulin to take based on recent food consumption and recent blood glucose readings, among other factors.

Incorporate automatic feeding mechanism

Principle
Machines performing powerful actions, such as cutting, grinding, and pulverizing, should not rely on users to manually insert materials.

Note: This principle has nothing to do with devices that feed cats when their owners are away for the weekend...

Machines that have a "feed" are often performing powerful actions, such as shaping, cutting, blending, pressing, grinding, and pulverizing. Saws, planers, wood chippers, printing presses, welders, and industrial mixers come to mind, all with moving mechanisms that could make short work of a body part or even a whole body. Standing near one of these machines could be hazardous in the event of a misstep or snagged piece of clothing or hair. Requiring users to manually feed material into such machines increases the likelihood of such accidents.

"According to OSHA, workers who operate and maintain machinery suffer approximately 18,000 amputations, lacerations, crushing injuries, abrasions and more than 800 deaths per year."[1]

Automatic feeding mechanisms stand to reduce the carnage.

Partially automatic vs. fully automatic feed mechanisms...

Partially **automatic** feed mechanisms require users to introduce material while standing a safe distance from the "business end" of the machine. For example, wood chippers with long conveyor belts enable users to load logs away from the chipping blades.

Fully **automatic** feed mechanisms automatically load material from a reservoir, hopper, or spool, enabling users to stay far away from moving parts. For example, commercial meat grinders automatically load meat into grinding, blending, and portioning mechanisms.

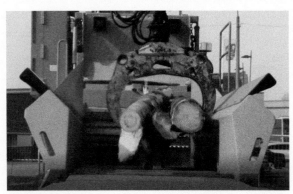

Photo: Colourbox.com

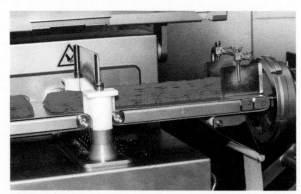

Photo: Colourbox.com

The future is here

Robots are already doing a lot of material feeding, such as placing a wooden or metal part into a jig ahead of a fabrication action, such as drilling. Like other machines, robotic systems should have safeguards that keep users and passersby in a safe position.[2] Physical barriers, such as enclosures, protect people from entering a robot's path and being struck. Similarly, sensors can be used to create an invisible barrier, immediately stopping the robot if a person enters a potentially dangerous area. And last (but not least), control panels and computers used to operate robotic systems can be located such that their operator remains outside the robot's range of motion.

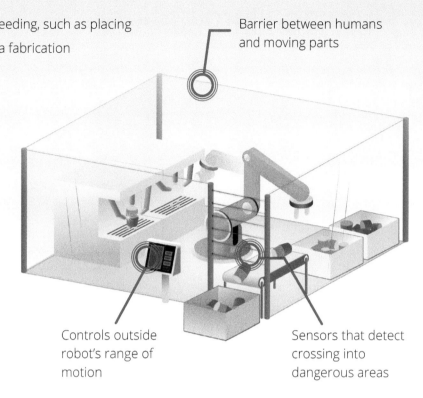

Barrier between humans and moving parts

Controls outside robot's range of motion

Sensors that detect crossing into dangerous areas

Additional precautions

Although worthwhile, safety signs warning users about placing their hands in harm's way are no replacement for an inherently safe mechanical solution that eliminates exposure to a hazard. When it is necessary for a human to feed material into a machine, and even for machines with automated feeding mechanisms, there are multiple ways to ensure that the workers' hands are not at risk.[3]

Decouple "Load" and "Start"
Require users to load material and then step away to activate the device from a safe distance, thereby ensuring the operator cannot start the machine while near moving parts.

(Photo: Colourbox.com)

Require safe hand position
Require the operator to press multiple buttons with both hands simultaneously to activate a machine, thereby ensuring that the user's hands are out of harm's way.

Employ sensors
Use a laser beam and sensor to deactivate a product if the beam is interrupted by the presence of a body part, or any object for that matter.

Limit free motion
Have the operator wear wristbands strung to a pullback device that restrains or pulls hands back if they cross into a hazardous location.

Provide illumination

Principle
Illuminate work surfaces, controls, and displays.

Show me the light

In some situations, illumination is essential to safe product use. For example, it helps to illuminate an environment to ensure safe passage. Lighting can also make people and objects more visible to others in ways that help prevent accidents. Integrating lighting into a product can also help people perform tasks more precisely, such as using an electric drill when one's hand is placed close to the spinning drill bit, or inserting a test strip into a glucose meter's illuminated slot and reading the test result on a back-lit screen. Below are additional examples of how lighting can be used to improve product safety:

Guiding lights
Lighting can guide users through an interaction or procedure. For example, the next button in a sequence could light up or blink, drawing users' attention.

Passive illumination
A reflective finish can be used to make objects (e.g., traffic cones) and people (e.g., runners or road workers wearing a reflective vest) more visible.

Reference lights
Lights can demarcate safe from unsafe areas and provide a point of reference to facilitate safe passage in dark environments.

Indicator lights
Indicator lights can be used to communicate system status or direct users when it is safe to proceed.

Head lamp
A head lamp provides task lighting while keeping the user's hands free.

Mounted lights
Lights can be attached (or mounted) to devices to illuminate the working area and make the device and user more visible to others.

Optimal lighting levels depend on the situation

Lighting should be sufficiently bright, but not blinding. Normal indoor light levels typically fall in the range of 200-500 lux.[1] Task light levels that fall in the range of 1,500-2,000 lux should be sufficient to enable detailed work.[2] Surgical lights produce extremely bright light in the 40,000 to 160,000 lux range.[3]

In some cases, automatic illumination in response to (1) a reduction in the ambient light level (e.g., sundown, normal lighting system power failure), or (2) motion, adds a measure of safety when people forget to activate lights or when their hands are occupied. Most modern cars have sensors that automatically recognize changes in the ambient light level and automatically turn on/off the

headlights. Similarly, dashboard controls and indicators often turn on automatically so they are visible at night. Some cars also flip the navigation screen from a daytime to nighttime display mode to avoid flooding the cabin with light that could interfere with the driver's vision.

Night Vision

Red lighting helps preserve night vision in cases where maintaining night vision is important, such as inside cars, airplanes and submarine control rooms. The eye is relatively insensitive to intensity changes in long-wavelength light, such as red light. As a result, the human eye can cope with relatively bright red light while remaining dark-adjusted.[4]

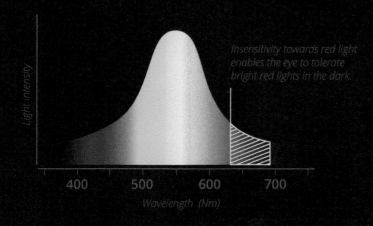

Insensitivity towards red light enables the eye to tolerate bright red lights in the dark.

Light intensity

400   500   600   700

Wavelength (Nm)

Add a "dead man's switch"

Principle
Products that require constant operator control should incorporate a switch that deactivates the device when the control is released.

What is a "dead man's switch"?
The term "dead man's switch" bluntly describes its purpose, which is to immediately shut down a function if the operator becomes incapacitated or is otherwise unable to continue actuating the function. According to some sources, the first use of a dead man's switch was in old streetcars. Drivers had to continuously hold down a handle while operating the streetcar; if they released the handle while the train was still moving, the motor would stop, the emergency brake would be applied, and the door lock would release.[1,2]

Different types of dead man's switches
We might not think of them as dead man's switches, but many products that could pose a danger if not actively controlled by an operator have functionally similar controls. Designers should carefully consider the scenario of use to determine which type of dead man's switch is appropriate for their product. Possible types of switches include:

Handle or pedal switch
Common in power tools and vehicles, users must continuously press handle-based switches (e.g., triggers) or pedals to activate potentially dangerous functions.

Touch-based sensor switch
Capacitive sensors built into a handle ensure a device is active only while the user's bare hand is in contact with handle.

Key switch
Some products require users to insert a key into the product to enable it. Sometimes, a lanyard or cord is attached to the key to enable a quick stop, such as the red cord attached to a treadmill's magnetic "key."

Seat switch
Some vehicles, such as forklifts, have seats with built-in dead man's switches that ensure the vehicle can only be operated when the there is a user in the driver's seat.

Location, location, location

It should be impossible for a dead man's switch to remain in the "active" position if the operator becomes incapacitated. For example, it should not be possible for a slumping, unconscious individual to hold a dead man's switch in the enabled position. Therefore, the switch's placement and activation style are important.

An effective dead man's switch should also be placed in a position that does not cause operator strain, either in terms of body position or activation force. After all, you do not want users developing creative ways to avoid having to manually press the dead man's switch, such as by taping it in the enabled position.

Worse   *Better*

It's not just good practice, it's the law

After an untold number of lawnmower accidents left people with fewer toes and worse, the US passed a law requiring lawnmowers to have dead man's switches (often referred to as "operator-presence devices"). Since 1982, lawnmowers must shut off, or at least rapidly stop the blade, if the operator releases his or her grip on the lawnmower's handle.[3] Personal watercraft have a similar function. If the rider falls off the craft, he or she will pull out an activation key and the craft will come to a quick stop–that is, if the rider is clipped to the key.

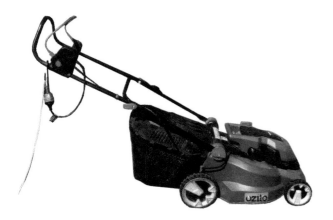

To run this lawnmower's engine, users must hold the red dead man's lever located above the lawnmower's handle. If a user releases the lever while the engine is running, the lawnmower will stop the blade.

Dead man's switches that kill (on purpose)

At the height of the Cold War's nuclear panic, the US Air Force's Strategic Air Command (SAC) division devised the Special Weapons Emergency Separation System (SWESS)–a dead man's switch intended for use in B-52 bombers carrying nuclear weapons. Unlike most dead man's switches, which disable a dangerous function when the user becomes incapacitated, SWESS was designed to detonate the on-board nuclear weapons if it determined that the plane and/or its crew became incapacitated.[4]

SWESS is an example of a fail-deadly (as opposed to fail-safe) strategy–a seemingly counterintuitive approach used in some military equipment, which aims to optimize overall safety by inflicting the most damage in response to a failure.[5]

If you're reading this sentence, then most likely no SWESS dead man's switches have ever been actuated overhead. Thank goodness for that.

Make it ergonomic

Principle
Ensure products match the intended users' physical characteristics and capabilities.

Ergonomics: helping prevent injury

Products reflecting poor ergonomics can injure people in immediate or cumulative ways. Examples of immediate injuries include tearing the rotator cuff (part of the shoulder) and slipping (i.e., herniating) a disc in the back, leading to a pinched nerve. Examples of cumulative injuries include carpal tunnel syndrome, which affects the wrist, and impingement syndrome, which affects the shoulder (see *Principle 100 – Minimize repetitive motion*).

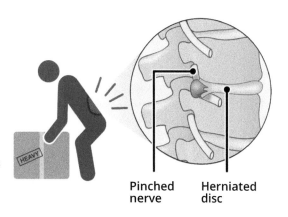

Pinched nerve   Herniated disc

A product has good ergonomics when its user interface—including all the "touchpoints"—accounts for the intended users' physical characteristics, capabilities, and potential disabilities.

An ergonomic computer workstation incorporates an adjustment mechanism that places the top of the computer's keyboard at a comfortable height—typically at elbow height when the user is seated with the elbows at 90° or slightly lower, and with the keyboard sloped downward and away from the user.[1] The goal is to enable users to reach the keys while maintaining a relatively "neutral" position—a position in which the back, neck, shoulders, and arms are relaxed.[2] It is sensible to complement an ergonomic workstation with an ergonomic chair that enables the user to adjust the lumbar support, arm rests, and seat pan tilt angles, and has other features that promote good posture, reduce strain, and simply make the chair comfortable.

Better

- Monitor height reduces neck and lower back strain
- Keyboard tray enables neutral wrist position
- Arm rests help maintain natural spine posture
- Adjustable seat enables feet to touch the ground

Worse

- Low monitor position increases risk of neck and lower back strain
- High keyboard produces contact stress and risk of carpal tunnel or tendinitis
- Raised seat produces pressure points / increased risk of nerve damage

Lift with your legs, not your back

Many injuries occur when people lift heavy objects. In the case of a work environment requiring someone to lift heavy objects to a storage shelf, the shelf should be positioned no higher than shoulder height, and perhaps lower, to avoid undue strain from repeated lifting (see *Principle 34 – Encourage safe lifting*). Similarly, a snow shovel that requires people to bend awkwardly to lift and remove heavy snow can be the number one enemy of a shoveler's back and arms. So, newer, ergonomic shovels incorporate bent shafts and additional handles to reduce strain. These innovative designs enable users to maintain a safe body position, as well as a neutral and powerful grip.

Let's talk about numbers

Good ergonomic design is informed by anthropometric data covering various dimensions of the human body, including those such as standing height, sitting elbow height, wrist breadth, and thigh circumference.[3]

anthropometrics:
an(t)-thrə-pə-'me-triks / *noun*

"Anthropometrics" comes from the Greek root words "anthropo" and "metron," meaning "human" and "measure," respectively.[4] *Or, in other words, "the measure of humans."*

In addition to body dimensions, anthropometric data can also relate to measures such as strength and range of motion of various body parts. The data is usually available for male and female adults. However, data is also available for special populations, such as children and pregnant women, as well as for people from specific regions of the world.

Textbooks, standards, and guidance documents provide abundant recommendations on how to apply anthropometric data to accommodate intended users. For example, there is guidance suggesting that the center of a computer display should be placed so that users can view it with a line of sight that is about 10 degrees below horizontal to avoid neck and eye strain.[5]

Ergonomic design is partly a numbers game. In an attempt to accommodate as wide a user population as possible, a manufacturer can design a product to fit persons ranging from the 5th percentile female to the 95th percentile male. This range will actually accommodate about 90% of adults.[6] Accommodating children and people with physical disabilities is another matter.

As an example, to ensure at least 95% of people can reach items on the top shelf of a book case, a furniture designer would want to ensure the top shelf is below the standing overhead reach for a 5th percentile female—73 inches off the ground.[6] Accommodating such a wide range of people reduces the likelihood that users strain their body or perform an unsafe action (e.g., standing on the first shelf) while trying to reach for an item on the top shelf.

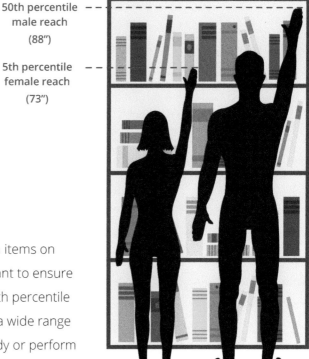

50th percentile
male reach
(88")

5th percentile
female reach
(73")

Armor it

Principle
Add armor in one form or another to deflect or absorb physical impacts and other forms of directed energy.

Determining when armor is needed

Products that are subject to powerful and potentially damaging impacts might require armor. Armor can take the form of a beefed-up outer case—call it an exoskeleton— or an accessory such as a Kevlar® wrap that offers protection against a physical assault.[2] Police officers often wear Kevlar vests for personal protection, particularly against gunshots. So do police dogs![3] But, the material has many more purposes, such as protecting bicycle tires against punctures, strengthening helmets against impacts, and helping canoes and kayaks resist damage (and the possibility of sinking) due to chance meetings with river boulders. Carbon fiber and other materials can serve some of the same purposes without adding significant weight to a product.

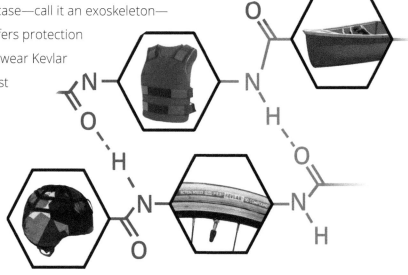

Kevlar is a polymer receiving its strength from the hydrogen bonds between polymer chains.[1] It is used in a variety of applications, including bullet-proof vests, canoes, bike tires, and helmets.

Shielding tubes and wires

The integrity of wires and tubes can depend on conduits and coverings that might also be considered a form of armor. Such armoring can prevent shocks and fires that could otherwise occur if an unprotected wire was chafed by a sharp-edged object or punctured by a drill bit, screw, or nail, for example. In the case of fluid tube protection, a braided metal covering can help protect against punctures and also prevent bursting due to overpressurization. Accordingly, braided metal wraps can help prevent the accidental release of hazards such as toxic chemicals, hot fluids, and steam.

Photo: Colourbox.com

Armored electrical wire
Rigid metal conduits are often used in commercial applications to protect electrical wires.

Fluid lines
High-pressure fluid lines can be run inside of braided metal coverings (i.e., conduits) or the equivalent to prevent bursting.

Today's high-tech tanks employ multiple forms of armor including steel plates and ceramic blocks (think Space Shuttle tiles), which serve to resist both impact and heat.[4]

Some mobile phones, laptops, and tablet computers are toughened by the addition of strong cases, such as those made of magnesium alloy,[5] thereby making them MIL-STD-810G compliant[6] and less vulnerable to damage if dropped.

Ruggedizing field equipment to improve safety

Displays can be toughened as well (see *Principle 7 - Temper the glass*). A tough computer might not directly save lives in the same manner as a helmet, but its ability to continue functioning could be essential to life-critical communications, such as those on a battlefield or in a search-and-rescue scenario. Similarly, a protective case on a cell phone might protect a cell phone belonging to a hiker lost in the cold, enabling him or her to make an emergency phone call.

Protecting against radiation

The concept of armoring can be extended to materials that protect people against radiation. A common solution is lead shielding, which is used extensively in medical facilities to protect people against X-rays. Dental patients and individuals working in catheterization labs that perform fluoroscopy usually don lead-lined clothing (e.g., vests, aprons, collars) that shields against the penetration of ionizing radiation. Notably, users should always be made aware of radiation-related hazards (see *Principle 72 - Indicate radiation exposure*) so they can take any necessary additional precautions beyond physical shielding.

Types of radiation and their penetration strength[7]

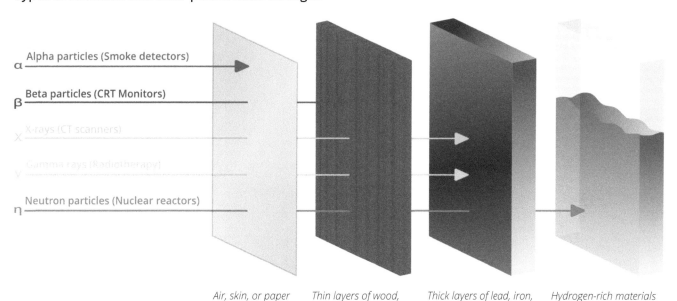

α — Alpha particles (Smoke detectors)

β — Beta particles (CRT Monitors)

X — X-rays (CT scanners)

γ — Gamma rays (Radiotherapy)

η — Neutron particles (Nuclear reactors)

Air, skin, or paper   Thin layers of wood, plastic, or aluminum   Thick layers of lead, iron, or other metal plates   Hydrogen-rich materials such as water or concrete

Incorporate a lockout mechanism

Principle
Incorporate a lockout or interlock mechanism to prevent untimely and potentially harmful interactions.

The key to safe design (pun intended)
The temperature inside a self-cleaning oven can reach 1,000°F during the self-cleaning process.[1] Also, oven compartments can fill with potentially toxic fumes during the cleaning cycle. That is why ovens have door locks that stay engaged during the entire cleaning cycle (i.e., to prevent burns and inhalation of toxins).

The terms "interlock" and "lockout" are sometimes used interchangeably. However, there is a subtle difference between the two. A "lockout" usually refers to a protective mechanism that cannot be overridden during a phase of product use. By comparison, an "interlock" usually refers to the requirement to perform two actions simultaneously or in series to get a result.

Lockout	Interlock
An oven's door locks during the self-cleaning cycle to prevent users from opening it and exposing themselves to extreme heat and fumes.	*A kitchen lighter requires users to perform two actions—sliding the lever and then pulling the trigger—to produce a flame, thereby adding a measure of protection against accidental activation.*

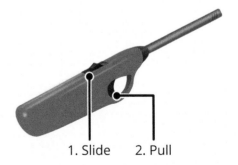

1. Slide   2. Pull

The quintessential lockout is, of course, a lock. Locks can be added to many different types of products to prevent unauthorized use, as well as to protect people from something hazardous. For example, some snowblowers cannot be used without inserting a key to unlock and start the snowblower. Parents can simply remove and hide the key and don't have to worry about children turning on the hazardous machine and trying to play with it like a toy.

Examples of lockouts

Lockouts are a common feature of amusement park rides, such as roller coasters. It is common for restraints to lock into place and stay that way until an attendant or automated system releases them once the ride is over. Ensuring that the restraints are properly sized and configured is a separate challenge (see *Principle 95 – Provide Restraints*).

A breathalyzer linked to an automobile's ignition system is an example of a lockout, preventing the engine from starting if the operator has consumed too much alcohol according to the legal limit. However, if the system determines that the operator's blood alcohol content is below the legal limit, the engine will start, and the driver can be on his/her way.

Front-loading washers and dryers have doors that automatically lock when their drums are in motion. The lockout prevents ill-advised attempts to add or remove items when the moving drum could cause harm or, in the case of the washer, an escaping wave of hot, detergent-laden water could cause skin and eye injuries.

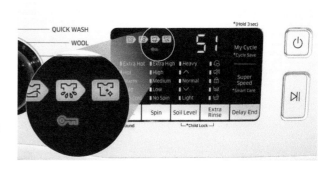

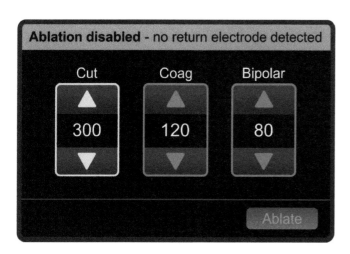

Some software applications also have lockouts that prevent users from taking action when it is not safe to do so. For example, some electrosurgical systems (i.e., medical devices that use energy to remove body tissue) have grounding pads that operators must place on the patient's body to prevent electrical surges.[2] However, if the system does not detect contact between the grounding pads and skin, the on-screen ablation control is disabled (e.g., "grayed out"), thereby reducing the risk of burns.

Eliminate or limit toxic fumes

Principle
Use ventilation and filtration to eliminate or limit toxic fumes.

Clear the air

With some exceptions, people generally recognize toxic fumes when they see or smell them. Have you ever held your breath when driving past a vehicle emitting blue or black smoke? Some automobiles can recirculate inside air to prevent the intake of contaminated air from the outside, and are equipped with filters (e.g., High-Efficiency-Particulate Air (or HEPA) filters) to cleanse the air before it passes through the ventilation system to the cabin. However, for the benefit of all, the far better solution is to reduce or eliminate the toxicity of exhaust gases at their source. For example, electric and hydrogen-powered vehicles emit no tailpipe pollutants (although their energy production plants might).

Did you say photocopier?

Unfortunately, there are many more products in the environment that can produce noxious fumes. A photocopier is a product that many people might not normally think of as polluting. But, photocopiers can release a medley of toner particles (e.g., carbon black), resins or polymers, and petrochemicals (e.g., naphtha) into an office space,[1] which can be dangerous in a room that lacks sufficient ventilation. In effect, a photocopier can produce ozone (i.e., smog), and, in turn, ozone can reportedly irritate the eyes, upper respiratory tract (i.e., nose, throat and airways), and lungs. Other symptoms include headaches, shortness of breath, dizziness, general fatigue, and temporary loss of olfactory sensation.[2] These reported hazards can be mitigated by the installation of an activated carbon filter.

The size of the room where the ozone-generation photocopier is located can contribute to ozone levels. In one study, a machine run in a 350-square foot room generated ozone counts in the range of 500 to 800 parts per billion—ten times the allowable limit.[3]

Tips to reduce risk of toxic fumes

Provide reminders to maintain the air filtration system and change filters regularly (see *Principle 15 – Provide Reminders*).

Make it obvious when toxic fumes are present by adding color or an odor (i.e., distinct smell) to fumes so leaks are more easily detected.

Design the filter to change color with usage to highlight when the filtration system requires maintenance.

Include a detection system that alarms when leaks are detected or when fumes reach an unsafe level.

Seal connections and passageways to prevent toxic fumes from escaping.

Wreaking haVOC

In the past, certain paints, wood finishes, waxes, cleaning supplies, pesticides, moth repellents, and many more products have been a significant source of volatile organic compounds (VOCs) linked to the following ill effects (amongst others, including cancer in animals):

- Damage to the kidneys, liver, and central nervous system
- Irritation to the skin, nose, throat, and eyes
- Headaches, nausea, loss of coordination, and fatigue[4]

Today, government regulations and industry action have led to the development of low and zero VOC options for many of the products listed above. For example, in the US, the Environmental Protection Agency (EPA) limits the number of VOCs that coatings manufacturers can have in their products.[5]

 Did you know that natural gas is odorless? An odorizer (typically mercaptan, a harmless chemical that smells like sulfur) is added to make it easier for people to detect natural gas leaks.[6]

Anesthetic gas scavenging systems are designed to remove anesthetic gas from environments, such as operating rooms.

Example: Anesthetic Gas Scavenging Systems

Hospital operating rooms have been shown to contain traces of waste anesthetic gases—the result of recovering patients exhaling anesthetic gas, and leaking machine connections, tubing, and masks.[7] Anesthesia gas scavenging systems, absorbers, and ventilation systems can reduce operating room workers' exposure to waste anesthetic gases.[8]

Add a horn, whistle, beeper, or siren

Principle

Use a horn, whistle, beeper, or siren to draw users' attention or warn users about a potentially dangerous situation.

Horns as a warning indicator

There are many types of horns. Rams grow one type. Musicians play another type (e.g., French horn, trumpet, tuba) that traces back to blowing through a hollowed-out ram's horn or conch shell. Automobiles are equipped with a third type of horn, which is intended to warn those in proximity. Similarly, some might consider a warning beeper to be the electronic equivalent of a horn, and a siren to be a horn that produces a continuous sound.

Products can be equipped with horns for decorative reasons, but, arguably, their primary purpose is to issue a warning: "Watch out!"; "I'm here!"; "Get out of the way!"; "Alert!". In the case of an automobile, it might be all four.

Conceivably, a horn is a necessary safety feature of many products because there might be an immediate need to draw attention to avoid an accident. As an example, bicycles should probably have loud horns, but most only have a small bell that rings in a cute but ineffective manner when up against a truck that is drifting toward the shoulder and bicyclist. That is why some cyclists now equip their bikes with air-horns that use compressed air to create a loud beep (up to 115 dB—equivalent to a car's horn) to alert motorists. Sirens take horns and beepers to the logical extreme. The attention-getting noise doesn't stop and can typically change in frequency. Applications include ambulances, fire trucks, and severe weather (e.g., tornado) alert systems.

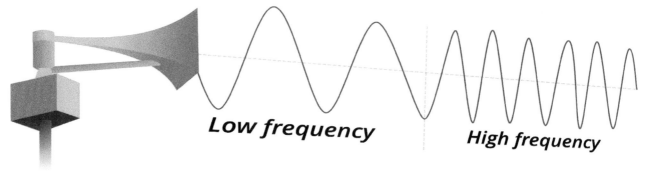

Low frequency

High frequency

Sirens often oscillate between low and high frequencies to overcome environmental noises that might otherwise mask the siren's sound.[1]

Backup horns

For decades now in the US, backup horns (i.e., backup beepers) have typically been installed on heavy equipment, such as front-end loaders and forklifts.[2] Their repeating sound might seem monotonous and habituating, but they are difficult to ignore and are intended to create a safer environment. One might consider them the offspring of whistles that alert people when a train is approaching.

The US Occupational Safety and Health Administration (OSHA) regulations states, "No employer shall use any motor vehicle equipment having an obstructed view to the rear unless: The vehicle has a reverse signal alarm audible above the surrounding noise level or [t]he vehicle is backed up only when an observer signals that it is safe to do so."[3]

Make some noise!

To draw attention in a safety-critical scenario, horns, beepers, and sirens need to stand out against background noise. This can be accomplished by setting their tonal frequency to a level unlike the expected background noises, such as by making them higher in tone (e.g., 500-1000 Hz).[4] Another way is to make them louder than 65 dB and at least 15 dB above the steady ambient noise level.[5]

Examples of sounds and their intensity levels[6]

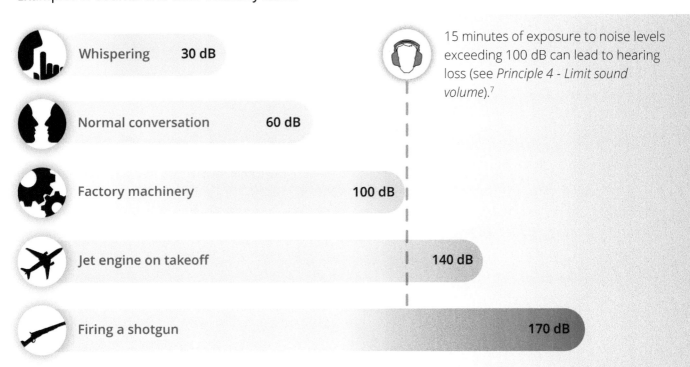

Whispering	30 dB
Normal conversation	60 dB
Factory machinery	100 dB
Jet engine on takeoff	140 dB
Firing a shotgun	170 dB

15 minutes of exposure to noise levels exceeding 100 dB can lead to hearing loss (see *Principle 4 - Limit sound volume*).[7]

74

Let users set the pace

Principle
When possible, a product should give users control over the pace at which they perform tasks.

Why let users set the pace?
When performing a task, people tend to prefer when they can set the pace, rather than having to match a pace set by a product. In this context, the term "product" encompasses devices, machines, or systems.[1] By giving people control, they are more likely to perform mental and physical tasks more effectively. Imposing a specific pace is problematic because different people naturally perform tasks at different speeds.[2] If a user cannot keep up with the product's pace, it can jeopardize the task outcome and even create a safety problem. For example, fast-paced assembly line work can lead to severe injuries, such as a crushed limb, that might occur when a worker cannot keep up with the artificially-set pace of assembly line machinery and physical conflicts develop.[3] Is anyone else picturing Charlie Chaplin cruising down the conveyor belt in the infamous film *Modern Times*?

Going up...
A simple example of an artificially-paced (often called machine-paced) task is getting on an elevator. When the bell rings, the elevator door opens, and there is a limited amount of time to step into it before the door closes again. Missing the window of opportunity to ride an elevator might not be a safety concern. Nor is stepping in right when the doors start to close, because there is a sensor that prevents the doors from crushing the body (see *Principle 56 - Use sensors*). But, elevator navigation was not always as safe. Some early elevators did not even stop at each floor. One type in particular, called a paternoster, was popular in the early to mid-1900s.[4] A paternoster's compartments (i.e., cabs) move continuously, and riders need to step in and out of the cabs at the right instant; otherwise, they could fall into an increasingly deep pit or get hurt stepping up into a high and still-rising cab.

Someone stepping into a paternoster's compartment (i.e., cab).

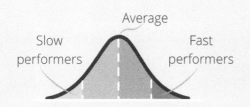

A bell curve describes the variability in time that a sample of people, who are trying their best, might require to complete a task. Limit the task time too much and people reflected in the tail of the distribution—the slow performers—could feel stressed and make more mistakes. People reflected in the opposite tail of the curve—the fast performers—might feel hindered by a slow task.

Aggressive defenders

Another example of an artificially-paced task is mowing the grass with a single speed, self-propelled mower. Such a mower relieves the user of having to push, but requires the user to match its motor-driven pace. While the person mowing has the option to let go of the mower, they might be inclined to hold on as the mower drives itself into a hazardous spot. One possible and unfortunate result is disturbing a bees' nest and what that can lead to—bee stings by aggressive defenders! Another mishap could be striking a rock that gets thrown toward the user or a bystander. Today, auto-stop controls (see *Principle 23 - Add a "dead man's switch"*) and adjustable-speed mowers lower this risk. Some have special transmissions that make the mower drive forward at precisely the speed the user chooses to walk.[5]

Worse

Better

Adjustable-speed mowers let users set the speed at which they choose to walk to ensure users keep up with the pace of the mower to prevent hazardous situations. (Photo courtesy of American Honda Motor Co., Inc.)

Letting users pace themselves enables them to perceive a situation, consider alternative actions, choose one, and confirm its effectiveness. Constantly or occasionally rushing someone to complete a task can invite failures (i.e., mistakes).

 Proposed in 1952, Hick's Law states that people faced with many choices will take longer to make their choice; reaction time increases logarithmically based on the number of choices.[6] So, the more information a user needs to process, the more time the user will need to make a well-balanced decision.

How to let users set the pace

When a product has to move things along to function properly, you can still help users keep up with the pace.

- **Timer**
  Display a countdown timer so that users have a clear sense of how soon they must complete their task.

- **Speed Control**
  Enable users to set the product's speed (e.g., low, medium, high) to a comfortable pace.

- **Alert**
  Although the tactic can cause stress, tell users when they need to go faster to meet a target task time.

Inactivity alert
Changes will be discarded in **30 sec**

- **Replay**
  Enable users to replay a video or animation if they did not pay attention or need to watch it again to understand the message.

- **Pause**
  Enable users to pause a process if necessary, perhaps to catch up with the action or correct a mistake.

Protect against roll-over and tip-over

Principle

When devices could crush or injure a person if they tip or roll over, incorporate features that will protect the user and/or employ preventative measures.

Protecting against roll-overs

When there is a potential for equipment carrying a person to roll over, it probably needs a roll-over protection structure—referred to as ROPS in many industries. In fact, the law requires it in some cases, such as with certain agricultural tractors and material-handling equipment in the United States (per regulations enforced by OSHA). ROPS work best when there is also a seat belt that holds the operator inside the protected area and prevents her or him from being ejected and potentially crushed during a roll-over accident.

> *"The tractor is the leading cause of death on the farm....The use of ROPS and a seat belt is estimated to be 99% effective in preventing death or serious injury in the event of a tractor rollover."* [1]

The National Highway Traffic Safety Administration's (NHTSA) 2009 rule says "that vehicles weighing 6,000 pounds or less must be able to withstand a force equal to three times their weight applied alternately to the left and right sides of the roof." [2] Many sports cars and off-road vehicles also have a ROPS. Indeed, a ROPS is what gives some of these vehicles a tough look that some owners seek. In other cases, the ROPS has a more subtle appearance. Consider modern convertibles that have built-in roll protection (e.g., extending pillars, hoops) that stay hidden until they are needed. If the vehicle's on-board sensor indicates a roll-over is in progress, the protection device extends or pivots into place, sometimes in less than a third of a second. [3]

With ROPS — Roll bar positioned over driver's head.

Seat belt use is also critical.

Photo: Colourbox.com

Roll-over safe zone

Without ROPS

Injury due to lack of roll-over protection

Similar to adult-sized vehicles, bike trailers can be designed with internal structures that offer some protection to the child inside. Children's bike seats, in combination with helmets, offer protection against injury due to bicycle tip-overs.

Even if it doesn't roll over, it can still tip over

One might associate the term "roll-over" with moving vehicles, but people might encounter similar hazardous events with stationary objects as well. For example leaning on or moving a large appliance might cause the product to tip over and crush a bystander (or the mover). Medical equipment designers can find specific guidance on preventing tip-overs in the standard IEC 60601-1 Edition 3.1, which describes how to test a medical cart's stability when positioned on a typical horizontal surface during normal use.[4] Moreover, the standard requires that medical carts not tip over while in a normal use position if on a 5° incline and not tip over while in a transport position (e.g., doors shut, arms set in) on a 10° incline.[5]

Children and tip-over hazards

According to the US Consumer Product Safety Commission (CPSC), "1 child dies every 2 weeks when a TV, furniture, or appliance falls on him or her."[6] Adding to that frightening statistic, performance testing of dressers and chests by the not-for-profit group Kids in Danger (KID) revealed that "Only 9 of the 19 units [that KID tested] passed performance tests based on the current tip-over safety standard, ASTM F2057".[7] In response, KID suggested, among other things, that the CPSC expand its educational efforts, and that manufacturers develop restraints (e.g., anchors that attach to a wall) that are easier to attach and do not require tools. KID also identified desirable stabilizing features from a few units that tested well:

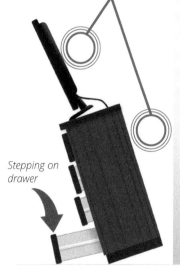

No restraints!

Stepping on drawer

When effective restraints are missing, a child stepping on a low drawer in an attempt to climb a dresser (e.g., to interact with a television on top of it) can lead to a deadly tip-over.[8]

- Interlocking drawers that only allow one drawer to open at a time

- A wider storage bin at the bottom of the unit that reduces potential for tipping

- A heavy panel across the bottom half of the back that acts as a counterbalance

Moose tests and fishhooks

Many vehicles are available with some form of electronic stability control (ESC) that can intervene when vehicle movements suggest a loss of control, such as a roll-over, is imminent. Some tractor trailers have ESC features that cut the throttle and apply the brakes in ways that help restabilize the vehicle if a potential roll-over is detected.[9]

A vehicle's roll-over potential can be evaluated using the Static Stability Factor: a formula comparing a vehicle's track width (distance between the center of the two wheels on an axle) and center of gravity height. As you might guess, vehicles with a high center of gravity (like SUVs) are more likely to roll over.[10]

A vehicle's roll-over potential can also be evaluated under dynamic conditions, using a "fishhook maneuver" or "moose test." During these tests, a driver swerves the vehicle suddenly in one direction and then again in the opposite direction, mimicking a driver trying to avoid an obstacle like a large moose, to see if the vehicle tips up on two wheels and/or remains in control.[11,12]

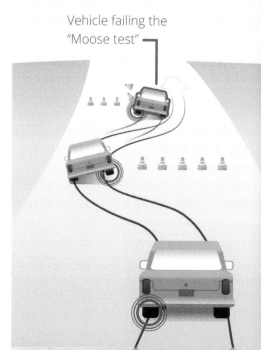

Vehicle failing the "Moose test"

Exemplar 3
Tractor

Tractors of various size and purpose are hardworking machines loaded with features intended to keep operators safe in challenging work environments.

Protect against roll-over and tip-over

Principle 30 - pg. 77

Roll-over protection structure (ROPS), which frames the cabin, protects the operator from being crushed by the tractor if it tips over.

Let users set the pace

Principle 29 - pg. 75

Throttle and cruise control mechanisms enable operators to adjust the tractor's speed as needed to suit the task at hand.

Provide illumination

Principle 22 - pg. 61

Lights make the tractor visible to others and illuminate the path ahead to facilitate safe operation in dim lighting.

Make it ergonomic

Principle 24 - pg. 65

Adjustable seat enables comfortable posture for operators of different sizes. Driver controls are also positioned within reach to help reduce arm and back strain.

Eliminate pinch points

Principle 98 - pg. 227

Automatic hitch "self-attaches" to accessories (e.g., mowers, backhoes), eliminating the need for manual attachment and reducing the risk of a pinch or injury during attachment.

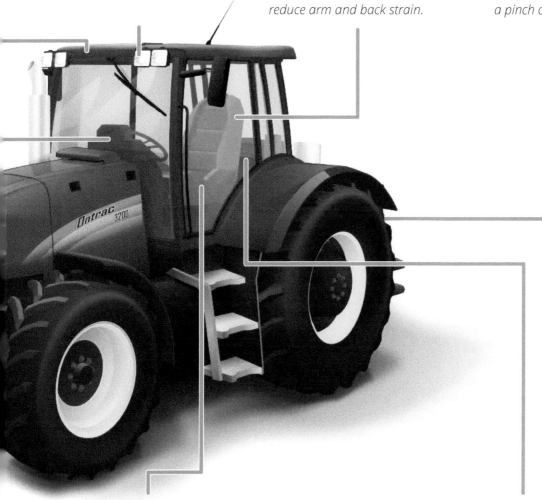

Add a "dead man's switch" / Use sensors

Principle 23 - pg. 63 / Principle 56 - pg. 135

Engine shuts off automatically when the sensors no longer detect the driver in the driver's seat, thereby preventing a "runaway" if the driver falls out of the vehicle.

Armor it / Eliminate or limit toxic fumes

Principle 25 - pg. 67 / Principle 27 - pg. 71

Sealed cabin protects the operator from outdoor heat and cold, flying debris, and engine fumes. The cabin's interior air filters also help provide healthier, breathable air for operators.

Make parts move, deform, or disconnect

Principle

To protect users from sudden impacts, sharp edges, and entanglement, ensure products fail safely by moving, deforming, disconnecting, or breaking away.

Why make products purposefully "break"?

Users could be injured if a product fails in an unsafe manner. For example, if a parked car's side mirror were designed to stay put when hit, a pedestrian or cyclist striking the mirror could suffer blunt trauma. Fortunately, side mirrors are designed to pivot (or break away) when struck, thereby reducing the chance of significant harm. The lesson is clear: designers can engineer certain parts to move, deform, disconnect, or even break away in reaction to excess or sudden force. The goal is to make such parts stable (i.e., resist force) up to the point that they would otherwise cause injury, and then give way in as safe a manner as possible.

Make parts move

Basketball hoops are designed to flex and then pivot on a strong, spring-supported hinge when a player dunks a ball and perhaps deliberately hangs off the rim. This mechanism is known as the "breakaway rim" and was invented by Arthur Ehrat in 1976.[1] It helps prevent shattering the backboard and causing glass to rain down. If the glass does break, the use of tempered or annealed glass can cause it to break safely into many small crumbles that are less likely to cause injury (see *Principle 7 - Temper the glass*). Another advantage of the breakaway rim flexing is the decreased impact on a player's wrist when he or she hits or holds the hoop.

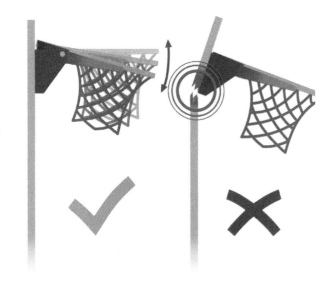

Snow blowers incorporate "shear pins" that break so that their turning auger (i.e., screw-shaped blade) will not tear through a solid object. This protects the machine from damage and can potentially save a person's entrapped limb. When a shear pin gives way, the auger disconnects mechanically from the powered shaft that turns it, and the auger stops.[2]

Make parts deform

Deformation effectively reduces impact forces by absorbing and spreading them over a larger area. That is the principle behind bike helmets, which are designed to sacrificially deform and even break apart if the user's head strikes a hard object (e.g., ground, curb, telephone pole).[3]

Impact forces can also be reduced by designing stretch into materials and adding padding (see *Principle 5 - Include pads*). Dynamic climbing ropes and bungee cords slow down a fall by absorbing energy over distance and time, converting a potentially fatal drop into an adrenaline rush.[4]

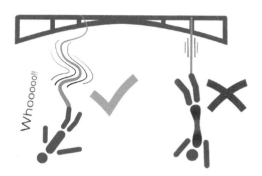

Make parts break away

Some computers, such as MacBook Pros, have a power cable held in place by a strong magnet. Snagging or tripping over the power cord usually causes the cable to disconnect, which can prevent a harmful tripping accident or prevent the computer from falling onto an unsuspecting foot (resulting in a damaged computer and a hurt foot).

Similarly, lacking a safety release, a lanyard (i.e., neck strap) that becomes entangled in rotating equipment could cause suffocation, neck injury, or even decapitation. On the other hand, a breakaway safety release would separate the lanyard at the breakaway point and prevent injury to the wearer.

Provide a handrail

Principle
Add a handrail to stairs and platforms to ensure people can move about or stand in place in a stable manner.

Giving users a helping handrail

Handrails are very...handy in terms of preventing falls. Perhaps the most common examples are handrails built into residential shower stalls and staircases. Some handrails assist individuals who have impaired strength, coordination, and/or balance. In fact, many countries have laws requiring handrails to be available in public facilities to make the facilities accessible to people with disabilities.[1]

Handrails and universal design

"Universal design (UD) is an approach to design that increases the potential for developing a better quality of life for a wide range of individuals. ... It creates products, systems, and environments to be as usable as possible by as many people as possible regardless of age, ability, or situation."[2] Handrails found on equipment, as well as architectural elements (e.g., ramps), are good examples of universal design, whereby able-bodied individuals can also benefit from accommodations intended for people with impairments.

Handrails are common features on heavy equipment, such as backhoes and oil trucks that require operators to climb staircases to reach the cabs or to access components that require maintenance. These handrails provide stability during work as well as prevent falls. Handrails in subway cars do not prevent falls from dangerous heights, but they are needed to stabilize riders in the routine event of forward, backward, and lateral accelerations.

If you find yourself walking down the aisle of a commercial airplane, check out the lower edge of the overhead luggage compartments. Some of them have an integrated handrail to help passengers and crew members cope with frequent bouts of turbulence.

Some airplanes feature handrails integrated into the bottom of the luggage compartments.

Tips for optimizing handrails

To serve its function, a handrail must itself be accessible and readily gripped. Although handrails come in many shapes, round ones are a common breed. Rail diameters in the range of 1.25" to 2" are recommended to enable a secure hold with the fingers wrapping around the rail enough, but not too much that it interferes with a strong, palmar grip. According to some building codes pertaining to non-circular rails, the maximum cross-sectional length is 2.25", and the rail's perimeter must be between 4" and 6.25". But, what happens if the rail's perimeter is larger than 6.25", perhaps due to the material used for the railing? In these cases, the rail must include grooves on both sides for "graspability's" sake.[3]

Rails with perimeters > 6.25" must include grooves on both sides to afford grasping the rail

Round rails should have diameters between 1.25" - 2"

The Occupational Safety & Health Administration (OSHA) also outlines several requirements for guardrails and safety rails.[4] Here are a couple excerpts:

- **1910.29(b)(1).** The top edge height of top rails, or equivalent guardrail system members, are 42 inches (107 cm), plus or minus 3 inches (8 cm), above the walking-working surface.

- **1910.29(b)(3).** Guardrail systems are capable of withstanding, without failure, a force of at least 200 pounds (890 N) applied in a downward or outward direction within 2 inches (5 cm) of the top edge, at any point along the top rail.

A case study in universal design: The city of Toronto

In June 2000, the city of Toronto began an effort to make the entire city accessible, enabling locals and visitors alike to enjoy the city's public buildings, parks, and open spaces, regardless of any impairments or disabilities. In April 2004, the city published the City of Toronto Accessibility Design Guidelines, which contains hundreds of guidelines for the design of public spaces. For example, the guidelines suggest that the slope of a sidewalk leading to a bridge should be no more than 5%. Notably, the document provides dozens of guidelines on handrails. For example:[5]

- Continuous handrails should be provided on both sides of ramps or stairs, or wherever three or more steps are present.

- Handrails on viewing platforms or terraces should be designed to enable a seated person (e.g., a wheelchair user) to see under the handrail.

- The width between handrails on interior ramps should be no more than 1,100 mm (43.3") and no less than 900 mm (35.4").

Seems like Toronto knows a thing or two about handrails, eh?

Prevent entrapment

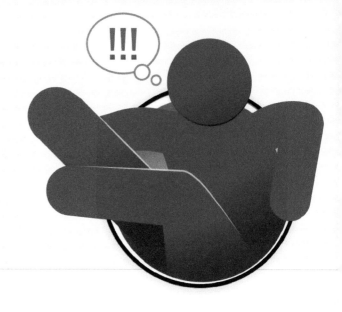

Principle
Design products to prevent users from becoming entrapped.

Entrapment is too common

The term entrapment might conjure scenes from 20th-century movies that show people sinking in quicksand. For what it's worth, people usually only sink to their waist and no farther because the human body is less dense than liquefied soils (colloid hydrogels).[1] Regardless, sinking into quicksand does not really factor into designing for safety. We are more concerned with preventing potential cases of mechanical impingement and forceful retention by man-made products, which can lead to severe harm, such as dismemberment or asphyxiation.

Crib rails
Historically, too many infants have gotten their heads entrapped in crib railings, resulting in contusions, or worse, strangulation. This sad fact drove the development of a standard for the spacing of crib rails. In the US, the maximum rail spacing, which is intended to prevent an infant from slipping her or his small head into the gap, is 2 3/8 inches.[2,3]

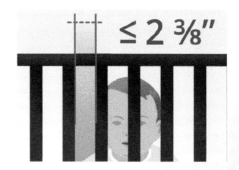

Structures
Entrapment in playthings and furniture is a major threat to child safety (see *Principle 63 - Childproof hazardous items*). This has led to many regulations on the mechanical design of such things as play structures. One requirement calls for component junctions to form angles greater than or equal to 55° to prevent a child from wedging their head or other body part between the two components.[4]

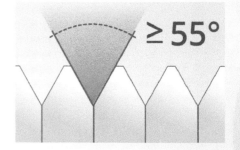

High-suction drains
Consumer Products Safety Commission statistics for the period 1999 to 2010 included 97 reported incidents of suction-related pool/spa/whirlpool bathtub entrapments including 82 injuries and 12 fatalities.[5] High-suction drains in pools and hot water spas now include covers that help prevent bathers from getting body parts (e.g., feet, hands) drawn into them so forcefully that it foils escape and causes drowning.[6] Such covers typically create adequate separation from the drain so that the level of suction at the cover's surface is low.[7]

When hospital beds make matters worse

Government and industry collaborators have established rules for the design of hospital bed guardrails in response to the multitude of accidents involving hospital patients—some of whom were unconscious to begin with—being injured or killed by the beds in which they were lying. Between 1985 and 2009, the US Food and Drug Administration (FDA) received reports of 480 deaths and 138 non-fatal injuries resulting from patients who became entangled, strangled, or trapped in beds containing handrails. There were also 185 incidents reported that did not result in injury because staff intervened in time. Most of the accident victims were elderly, frail, or confused.[8]

Guidance from the FDA includes the following protective measures:[8]

- Limiting or eliminating gaps between the mattress and bed structure

- Narrowing the openings between guardrails

- Employing patient positioning monitors (e.g., bed alarms) to detect when a patient is out of position and could become entrapped (see *Principle 18 - Predict hazardous situations*)

The red areas above highlight seven areas on a bed that have the potential to entrap patients. The detailed callout depicts neck entrapment at the end of a handrail.[9]

Insights

Neck entrapment in a product (e.g., playground equipment, machinery) can compress blood vessels (e.g., carotid arteries) dangerously, close an airway, and cause various musculoskeletal injuries. Additionally, being drawn into moving machinery—a rapid and progressive form of entrapment—has sadly been a routine cause of worker injury and death. For example, loose hair entrapment in rotating machinery has led to scalpings and deadly head injuries,[10] which explains the cautionary warnings found on many products or, in some cases, the requirement to wear a hairnet or use other means of hair containment.

Potential risk mitigations against entrapment in machinery include several safe design principles included throughout this book:

 Guards and/or automatic feeding mechanisms can keep users and their body parts away from dangerous moving parts (see *Principle 21 - Incorporate automatic feeding mechanism*).

 Breakaway components can release someone from a dangerous entrapment (see *Principle 31 - Make parts move, deform, or disconnect*).

 Automatic shutoffs can sense imminent or actual entrapment and shut off or disengage mechanisms before injury occurs or worsens (see *Principle 73 - Shut off automatically*).

  Well-designed warnings can inform users of risks, while also offering a modicum of liability protection against a common claim that the product manufacturer failed to warn (see *Principle 79 - Add conspicuous warnings*).

Encourage safe lifting

Principle
Products should facilitate safe lifting and inform users about safe lifting techniques.

How much can you handle?

Not surprisingly, people often incur musculoskeletal injuries when trying to lift something that is too heavy—in fact, lifting heavy items is one of the leading causes of injury in the workplace.[1] Injuries, such as to the spine, can be quite serious and lead to permanent disability. The chance of injury increases substantially when the lifted item's weight exceeds 51 lbs.

People have an easier time lifting heavy items that are in the so-called power zone, which is between the middle of a person's thighs and chest and close to the body.[2] Users can also avoid injuries by keeping their spine upright, bending their knees, and avoiding twisting.[1]

Such insights can guide designers to position handles, shelves, and storage compartments to facilitate safe lifting. That said, whenever possible, designers should aim to develop solutions that eliminate the need for lifting altogether, such as patient lifts that hoist patients out of their bed in hospitals and nursing homes.

Object in Danger Zone

Danger Zone

51 lbs

Power Zone

Danger Zone

Object in Danger Zor

When lifting heavy items:

- Keep spine upright.
- Do not twist spine.
- Do not bend or reach.

Patient lifts are used in many healthcare facilities to prevent healthcare professionals from injuring themselves while trying to lift patients.

Model photo: Colourbox.com

Safe lifting options

Product designers and manufacturers can help prevent lifting injuries by encouraging safe lifting in several ways:

- **Provide handles**
Place handles where they will make the product easiest to lift using the proper technique.

- **Show product's weight**
Clearly indicate a product's weight to dissuade someone from attempting to lift more than he or she can handle.

- **Encourage a two-person lift**
Indicate that one person should not attempt the lift alone—that he or she should get help.

- **Indicate safe lifting technique**
Include an instruction encouraging safe lifting technique.

Determining the potential for a harmful lift

The NIOSH (National Institute for Occupational Safety and Health) lifting equation is primarily used to calculate the maximum acceptable weight that a typically healthy individual could lift during an 8-hour work shift without increasing risk of lower back injuries. The NIOSH lifting equation has two outputs: (1) the Recommended Weight Limit (RWL), which answers the question, "Is this weight too heavy for the task?" and (2) the Lifting Index (LI), which answers the question, "How significant is the risk?"[3]

In practice, one would calculate the RWL and LI at both the origin of the lift (i.e., where an individual picks up the object) and the destination of the lift (i.e., where the individual places the object). If the maximum expected weight of the object is above the RWL, or the LI is above 1.0, it indicates the potential for injury at either the start or end of the lift. We won't go into much more detail about the NIOSH lifting equation (many other textbooks do that), but here are the primary factors that go into the equation (using the customary US measurement system):

LC = Load Constant represents the maximum recommended load weight for a lift under ideal conditions (51 lbs).

H = Horizontal location of the object relative to the body (inches).

V = Vertical location of the object relative to the floor (inches).

D = Distance the object moves vertically (inches).

A = Asymmetric angle or twisting needed to move the object (degrees).

F = Frequency and duration of lifting activity.

C = Coupling or quality of the user's grip on the object.

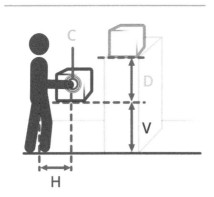

An overview of the NIOSH lifting equation's primary factors.

Make design features congruent

Principle

Products should conform to applicable design conventions, be internally consistent, and assume logical, intuitive forms.

Congruency supports design intent

Sirens shouldn't sound soothing. They should cut through background noise to draw attention, convey urgency, and perhaps even irritate the listeners to achieve the siren's purpose. As such, sirens are exactly what one expects them to be; their sound is congruent with their intent. It would be illogical to design a siren to sound relaxing, as if nothing were wrong, because people would be arguably less likely to notice and heed it.

"Congruent" means in agreement or harmony.[1] The elements of a product reflecting congruent design should work together to communicate the intended message.[2] Geometric congruence calls for similar items to have the same form and, purists might insist, the same size.

Consider a dialysis machine with blood tubes that carry blood from an artery and return it to a vein. The visually congruent solution is to color the arterial flow path component red and the venous flow path blue, noting that arterial blood is red and venous blood is darker red, but generally thought to be bluish (think cyanotic—a person turning blue). Adding congruence can make products more usable, but also prevent harmful use errors, such as connecting blood tubes to the wrong ports.

As in the previous example, color can be a powerful way to align perceptions with the associated physical properties and functions. A red LED indicator is a congruent way to indicate that a stove burner is on because people typically associate red with heat (e.g., a red glowing metal coil). A glowing red symbol that looks like a flame is arguably even more congruent.

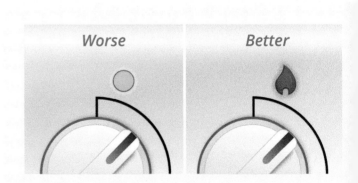

Worse   *Better*

☺ **BEWARE!** of incongruent design

The "Stroop Effect" is an increase in reaction time and error rate that occurs when one tries to say the text color of a word, rather than the actual word.[3] To experience the effect yourself, say the text color of the following words aloud:

Green   **Blue**   Red

Given the Stroop Effect, poor color choices in a user interface can cause users to misinterpret or overlook key information or controls. Consider how the following examples of color codes that conflict with design conventions might cause users to err:

 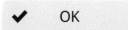
 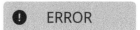

More examples of congruent design

"Not a step"
Unfortunately, some stepladders include a plank at the top of the ladder that looks just like another step but is not. Designers might try to make up for this incongruence with a warning label stating "Not a step." However, other ladder designs attempt to solve this incongruence with design features, such as:

- A plank across the ladder's top that is not shaped like another step

- A top *step* that is well below ladder's top and creates a larger platform

Mr. Yuk™
The Mr. Yuk™ symbol is a good example of congruent design. Children seeing the symbol on bottles should intuitively understand that the contents are bad for them because the cartoonish character looks sickened. The principle in play is not just to create a meaningful symbol, but also to stir a matching emotional reaction (see *Principle 93 - Label toxic substances*).

Photo: Colourbox.com

On the move

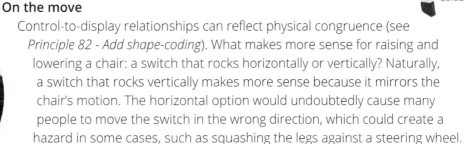

Control-to-display relationships can reflect physical congruence (see *Principle 82 - Add shape-coding*). What makes more sense for raising and lowering a chair: a switch that rocks horizontally or vertically? Naturally, a switch that rocks vertically makes more sense because it mirrors the chair's motion. The horizontal option would undoubtedly cause many people to move the switch in the wrong direction, which could create a hazard in some cases, such as squashing the legs against a steering wheel. Fortunately, the rocker switches that control some driver's seats are shaped and move similarly to the seat they actuate (e.g., image on left).

Provide visual access

Principle
Make sure critical information and key components are visible so users can monitor progress and make informed decisions.

Safety is in the eye of the beholder
Many things and processes need to be observed to help ensure safety. A good example is a patient undergoing a scan or radiotherapy. In the case of an MRI scan, technicians can observe the patient via a window between the scanner and control rooms. In the case of radiotherapy, visual contact is provided either via windows, or by video cameras and monitors, depending on precautions the technician must take to protect him/herself from radiation. Continuous visual contact reassures patients and enables technicians to ensure that their patients are correctly positioned for a safe diagnostic or therapeutic procedure.

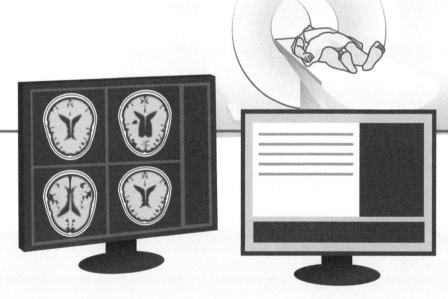

What could go wrong?
In the widely reported case of the Therac-25 radiotherapy machine, a young man seeking treatment for a tumor received a large overdose of radiation in part because the technician's video monitor was not functioning at the time. Had a video link been available, the technician might have seen and heard the patient reacting to the searing pain and stopped the therapy that led to a massive overdose and the patient's eventual death nearly five months later.[1]

A pilot's view through the head-up display on a Lockheed WC-130J aircraft.

Watch where you're flying

More and more commercial aircraft are incorporating head-up displays (HUDs), a flat sheet of glass positioned in front of the cockpit's front window that displays key operational information in the pilot's line of sight. The pilot can simultaneously look through the window and see key flight instrumentation without having to look down at the aircraft dashboard, allowing for increased situational awareness and better aircraft control.[2]

Some aircraft are currently outfitted with external cameras that enable pilots to see portions of their vessels that otherwise could not be seen through cockpit windows. The capability can help pilots view their landing gear and assess any external damage.[3]

 The Space Shuttle Columbia disaster might have been prevented if the crew had been able to inspect their shuttle's wings before re-entering the atmosphere. Damage to the left wing's leading edge was cited as the entry point for superheated air that melted the wing's internal structure.[4]

Many ways to see

Direct visual access or a camera view of a site might suffice to ensure safety. However, there might be a need for an enhanced or artificial image (e.g., ultrasound, infrared, MRI) to detect problems.

Some products incorporate access panels and holes that enable close-up inspections for problems, such as leaks and cracks. Holes, in particular, enable the introduction of scopes into remote locations requiring inspection. For example, safe, minimally-invasive surgery is enabled by video scopes that give surgeons the views necessary to operate. Advanced systems give surgeons a three-dimensional view of the surgical site, which might provide the necessary visual fidelity to operate effectively. Today's 3D systems require users to either wear head-mounted stereoscopic goggles or look into a special viewer.

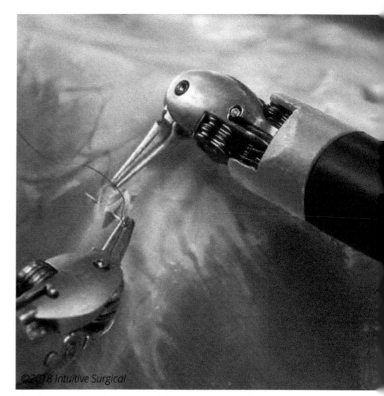

©2018 Intuitive Surgical

The da Vinci Surgical System enables a surgeon to perform minimally-invasive robotic surgery by sitting at a console and looking through a viewfinder that provides a highly-magnified 3D view of the surgical site.

Make glass panes visible

Principle

Make large pieces of glass more visible to prevent people from walking or running into them and sustaining blunt trauma or lacerations.

A crash course

The suggestion to make glass panes visible should not be confused with glass planes or gas pains. It's about making large pieces of glass—typically those forming walls and doors—more visible to people who otherwise might walk or run into them. The consequences of colliding with glass walls and doors can be serious lacerations caused by shattering glass shards. Another possibility is that someone who collides with a glass pane will suffer the same kind of injury as if he/she walked into a wall—the glass might not break, but facial contusions and fractures could result, for example. This has happened countless times in homes equipped with full-length glass storm doors. It also happens in public spaces and can result in lawsuits against businesses. For example, Apple was sued for $1 million by an elderly woman who suffered a broken nose when she reportedly walked into the all-glass front wall at an Apple Store.[1]

Someone running into a glass door with enough force could shatter the glass and be seriously injured.

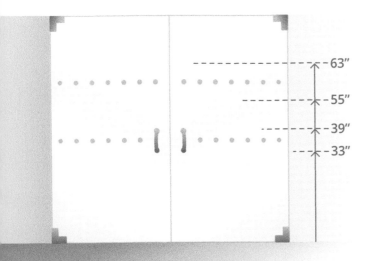

Easy as 33, 39, 55, 63

The risk of injury resulting from collision can be reduced by laminating or tempering glass (see *Principle 7 – Temper the glass*). It can also be reduced by making the glass more visible in some manner. In fact, there are standards in the UK that call for placing visible markers in two regions on full-height glass windows, partitions, or doors: one between 33" and 39" from the floor, and one between 55" and 63" from the floor.[2] Placing markers in these regions reportedly makes the glass visible for young people and old people, short people and tall people, and even pets.[3]

Methods for making clear glass more visible

Glass can be made more visible in various ways, many of which are surprisingly easy and inexpensive. Such methods include:

Acid etching the glass

Attaching a peel-and-apply film

Adding decals

Tinting the glass

Incorporating a contrasting frame

Using smaller, framed sections

Another protective strategy is to place items (e.g., a sofa, display table) in front of or behind large glass walls so that people see the obstacles if they do not see the glass.

No harm, no fowl

Making glass more visible to people is likely to help birds as well. A *New York Times* article from 2011 reported that: "New York is a major stopover for migratory birds on the Atlantic flyway, and an estimated 90,000 birds are killed by flying into buildings in New York City each year."[4] The article explained that after foraging for food in nearby parks, birds often collide with the lower levels of glass facades. According to some ornithologists and conservationists, these types of crashes lead to up to a billion deaths in the US each year and are the second-leading cause of death for migrating birds.[4]

In an effort to create bird-friendly buildings, Arnold Glass developed ORNILUX Bird Protection glass. The glass is coated with a UV-reflective coating that looks like a chaotic web of lines to birds but is virtually transparent to the human eye.[5]

Minimize distractions

Principle

Products should account for use in distracting environments and minimize distracting user interface elements.

Understanding distractions

Dealing with distractions can be considered a form of "multi-tasking," which, despite what the name implies, is typically the act of switching attention between tasks, rather than attending to many things simultaneously. Unfortunately, task switching often degrades performance because the more a competing task (e.g., a distraction) attracts our attention, even for a moment, the less attention we can apply to one primary task. As might be expected, research suggests that switching between tasks might increase the chance of error.[1]

Accordingly, you do not want to distract people performing critical tasks, such as an operator swinging a crane's arm above a populated construction site, a cytologist viewing a tissue sample through a microscope to determine if cancerous cells are present, or a pharmacist filling a prescription.

The "elephant in the room" or, rather, in the front seat

When discussing the consequences of distraction, accidents arising from distracted driving are arguably the "elephant in the room":

> *"[At] any given daylight moment across America, approximately 660,000 drivers are using cell phones or manipulating electronic devices while driving, a number that has held steady since 2010."[2]*

In fact, distracted driving has been widely chronicled as a growing cause of automobile accidents. According to the National Highway Traffic Safety Administration (NHTSA), there were 3,179 fatalities and an estimated 431,000 additional injuries involving distracted drivers in the US alone in 2014.[3] In response, almost all US states now ban texting while driving,[4] and Apple's iOS11 includes a "Do Not Disturb While Driving" mode that can mute incoming messages while the user is driving.[5]

Designers can mitigate distraction in myriad ways

Ergonomic controllers provide critical controls at the user's fingertips. (Photo: Colourbox.com)

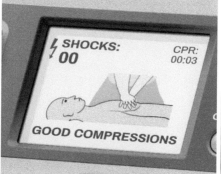

Simple prompts can be more effective during an emergency situation.

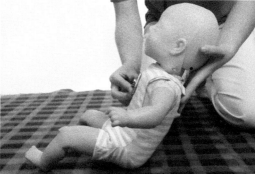

Proper training helps users learn safe and effective technique. (Model photo: Colourbox.com)

Require less effort

- Eliminate discomforts that can be distracting, such as having to stretch to reach a control or squeeze harder or more consciously to maintain a grip on an instrument.

- Make important information easy to acquire, which often calls for making the information reasonably large.

- Present information in an easily readable form (i.e., well-organized, prioritized, not overly dense).

Optimize for the situation

- Present only necessary information on a critical display, subordinate secondary information, and delete extraneous content.

- Automatically disable sources of distraction during critical phases of system operation.

- Do not overwhelm the user with too much sensory input (sounds, tactile cues, and visual signals) that might compete for attention rather than facilitate a task.[6]

Work with the user

- Provide training to help users remember proper technique and procedures when they conduct tasks in real-life scenarios (that can include distractions).

- When possible, let the user set the pace of tasks to help maintain high performance even in the presence of distractions (see *Principle 29 - Let the user set the pace*).

- Provide reminders and alerts that draw users' attention back to critical information and/or help users resume a task after attending to a distraction.

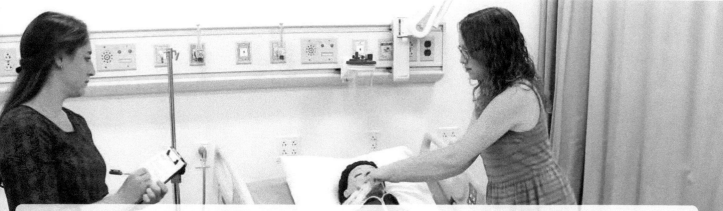

Don't forget to distract me!

Including potential distractions in usability testing is an invaluable way to test a device's ability to accommodate users in challenging environments. For example, including a phone call during a task with a home-use medical device can be a good technique because it a) redirects auditory, verbal, and visual attention, and b) pulls the test participant away from the original task for at least several seconds. Both of these effects are cognitive challenges that might undermine effective device use and reveal potential safety-related use errors.[7]

Prevent falls

Principle
Eliminate the cause of falls, or at least warn people about fall hazards so they can take precautions.

The dangers of heights

Falls from a standing position or a substantial height are a prominent cause of injury and death among children and adults. Infants can topple from changing tables. Adults can fall from ladders or platforms that lack a protective railing or equivalent. Seniors can fall due to muscle weakness or loss of balance. In the workplace, 15% of all accidental deaths in the US are fall-related, second only to transportation-related incidents.[1] Obviously, falls can result in minor injuries, such as bruising or abrasions. But, falls can also cause more serious injuries such as hip fractures, brain injuries, and death.[2]

 In the US, falls are the leading cause of injury and death in adults 65 and older. In 2017, the Centers for Disease Control (CDC) reported that 29 million falls led to 800,000 hospitalizations and cause 28,000 deaths each year.[3]

Fall detectors

A promising solution for elders who have balance problems that place them at greater risk for fall injuries is sensing mechanisms that employ accelerometers to detect falls and an on-product button to enable users to call for help. These devices are typically waterproof, so they can be worn in the shower. Fall detectors are often designed to be worn around the neck or wrist for easy manual activation in case of an emergency. Ideally, fall detectors automatically transmit an alarm in case the wearer loses consciousness or is too weak to press the button manually.

Mitigating fall risks

Below are several possible ways to prevent falls or reduce the severity of harm from a fall.

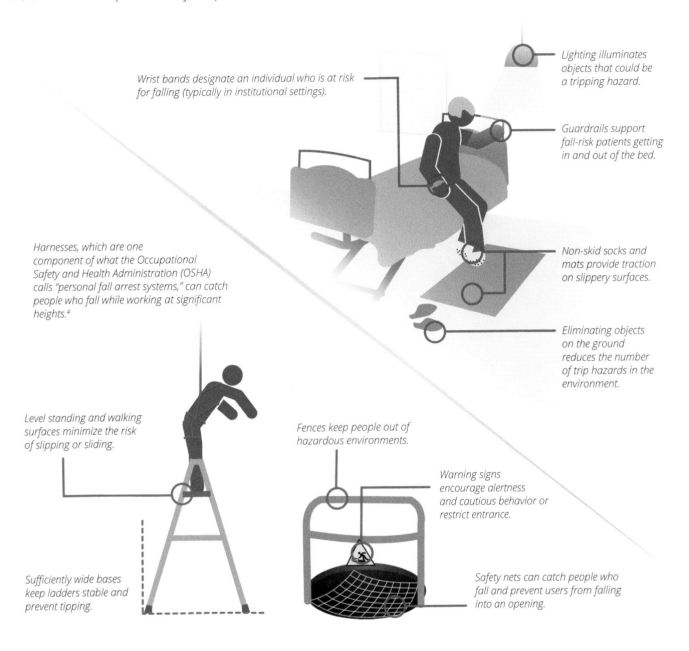

Wrist bands designate an individual who is at risk for falling (typically in institutional settings).

Lighting illuminates objects that could be a tripping hazard.

Guardrails support fall-risk patients getting in and out of the bed.

Harnesses, which are one component of what the Occupational Safety and Health Administration (OSHA) calls "personal fall arrest systems," can catch people who fall while working at significant heights.[4]

Non-skid socks and mats provide traction on slippery surfaces.

Eliminating objects on the ground reduces the number of trip hazards in the environment.

Level standing and walking surfaces minimize the risk of slipping or sliding.

Fences keep people out of hazardous environments.

Warning signs encourage alertness and cautious behavior or restrict entrance.

Sufficiently wide bases keep ladders stable and prevent tipping.

Safety nets can catch people who fall and prevent users from falling into an opening.

Safe crib height

Children's furniture and associated adjustments to it should be designed with a child's physical capabilities in mind. For example, a crib's mattress should be low enough that a child cannot climb up the crib's side and fall out. In fact, federal regulations in the US require there to be at least 26 inches between the top of the mattress foundation and the top of the crib side rail, and that crib mattresses have a maximum thickness of 6 inches (resulting in at least 20 inches between the top of the mattress and the side rail).[5]

MIN 26 in.

Make it slip resistant

Principle
Ensure that a product affords users a secure, slip-resistant grip.

What makes something slippery?

You probably consider it fun to slide across a smooth floor when wearing socks. But, what happens when you try to do the same while barefoot? Exactly—too much friction! But, without friction, we wouldn't be able to perform everyday activities, such as holding an object (it would slip right through our hands) or riding a bicycle (the tires would slide away from under us).

Due to relatively low friction, only a small push is needed to slide something across a smooth surface (e.g., hardwood floor).

Static friction is the force resisting two surfaces from starting to slide against each other. Dynamic friction is the drag force between two surfaces already sliding against each other. Depending on the combination of materials that contact each other, the amount of each type of friction varies. Some materials produce relatively high friction, like car tires on asphalt, while other materials produce lower friction, like skis on a snowy slope. The lower the friction between two materials, the easier it becomes for them to start and then continue to slide.

Due to relatively high friction, a much bigger push is needed to slide something across a rough surface (e.g., concrete).

Potential slipping injuries

There are all kinds of slips that could cause harm. Certainly, losing one's footing on a slippery floor can lead to injurious falls. That is why a lot of people place rubber mats on the shower's floor to add grip and help prevent slips. The Centers for Disease Control and Prevention (CDC) recently reported that falls each year by older individuals in the US lead to more than $31 billion in annual Medicare costs,[1] and that falls are a leading cause of traumatic brain injuries.[2]

Dropping a heavy object due to having a slippery grip on it could also lead to all kinds of unfortunate events, including broken toes or a damaged device. These are two of the arguably more obvious consequences of slippery conditions. But, consider the potential consequences of a slippery seat on a tractor, a slippery surgical instrument, or a slippery handle on a stroller.

Get a grip

When it comes to people "losing their grip," smooth surfaces are a common culprit, suggesting the need to roughen them up somehow.

One solution is to add texture, such as knurling or ridges on a tool's handle, but a popular alternative is overmolding with a high-friction material, such as thermoplastic elastomer, that can feel quite smooth but also provides a good grip. Sometimes, the answer is to add high-friction strips (e.g., anti-slip tape) to stairs and pathways, or apply a non-slip finish to a surface where people will tread.

Knurling on a dental tool's handle facilitates a secure grip.

A textured handrail prevents a user's hand from slipping off of it.

Another approach is to give an object a shape that helps keep hands, feet, and bottoms in place. For example, a RECARO race car seat is shaped to keep a race car driver from slipping sideways during high-G cornering.[3] Of course, the seat's suede-like Alcantara leather and seat belts are also a big help. Similarly, the RECARO child seats provide greater protection in the event of a side impact collision. Some bicycles' handles contain raised ridges (i.e., contoured nubs), which keep the cyclist's fingers and, therefore, hands in place.

A RECARO child seat is designed to provide optimal protection to children.

Raised ridges on a bicycle's handles help keep the rider's hands in place.

In the 18th century, a Dutch scientist, Pieter van Musschenbroek, invented the tribometer, which is used to measure the friction force between two surfaces.[4] Today, there are many other methods and devices used to measure surface texture and friction, as described in ISO 25178.[5]

Too much friction can also be a problem

When there is excess friction between an object and a body part, a "friction burn injury" can occur. Minor friction burns produce redness of the skin—like a rug burn—and more severe friction burns can lead to blisters and perhaps the loss of the top layer of skin.[6]

Motorcyclists can suffer "road rash"—a severe friction burn—if they lay down their motorcycle and skid across the rough road surface.

Exemplar 4
Stepladder

Stepladders are great for accessing high, otherwise difficult-to-reach locations, but they can also be hazardous if they do not remain stable during use. Stepladders have myriad safety features that ensure users win the battle against gravity and remain safe.

Make design features congruent

Principle 35 - pg. 89

Non-weight-bearing top bar is shaped differently than the ladder rungs and is perched high above the top rung, clearly distinguishing it as unsuitable for use as a step.

Protect against electric shock

Principle 75 - pg. 177

Non-conductive, fiberglass side rails reduce the risk of electric shock if the ladder comes into contact with live electrical components.

Provide stabilization / Prevent falls

Principle 1 - pg. 15 / Principle 39 - pg. 97

Wide "footprint" and fairly low center of gravity help keep the stepladder from tipping over during use. Flat steps provide a level standing surface that minimizes the risk of a fall.

Provide a handrail
Principle 32 - pg. 83

Hand grip and large steps help users steady their balance by enabling them to maintain three points of contact on the ladder at all times (i.e., two hands and one foot, or one hand and two feet)—an OSHA ladder safety recommendation.[1]

Make it slip resistant
Principle 40 - pg. 99

Textured, non-slip surfaces create good traction between the user's feet (or footwear) and the step. Rubberized pads on the ladder's feet also help keep the ladder in place during use.

Encourage safe lifting
Principle 34 - pg. 87

Wheels eliminate the need for users to carry the somewhat cumbersome and potentially heavy ladder. However, the wheels do not contact the ground until the ladder is purposely tipped for transport, ensuring the ladder remains steady while in use.

102

Display critical information continuously

Principle

Make sure users can always view the information necessary to make safe decisions and avoid potentially dangerous situations.

Always and forever (almost)

Information essential to safety should be displayed when and where it is needed. This principle pertains to the type of information that might be necessary to assess a potentially dangerous situation and take precautionary action. Or, the information might be necessary to prevent certain errors that could lead to harm.

Sure, people can turn a dial or click on an on-screen control to display critical information upon request. But, such extra steps take extra time, create work, and require vigilance—all hurdles to ensuring safety. Plus, the display of critical information helps people maintain situational awareness—an awareness of what is happening at a given moment.

Fuel tank and battery levels constitute critical information to some vehicle operators, such as helicopter pilots and powered wheelchair users, respectively. The information has obvious importance when levels drop too low. However, it can be quite important to know when a tank or battery is full and to be able to judge the consumption rate so the operator can plan ahead and avoid a potentially dangerous situation altogether. Running out of fuel in a helicopter can mean attempting an emergency landing in an unpowered, autorotation mode during an urgent search for a flat and open space.[1] Running out of electric power in a powered wheelchair can mean definite inconvenience and possible risk of harm depending on circumstances (e.g., if the wheelchair loses power when crossing a busy street).

Found on essentially every airplane, the "six pack" consists of six instruments needed by pilots at all times.[2]

In the case of electronic medical records (EMRs), critical information includes the patient's key information (e.g., name, date of birth), allergies, and medications. That is why many EMRs present such information on every screen pertaining to the patient. Lacking such information, a healthcare provider might inadvertently schedule medical care (e.g., a diagnostic test) based on incorrect patient information or administer a medication that could trigger an allergic response.

Worse	*Better*

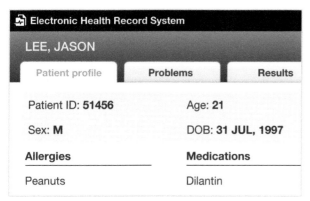

The patient's key information, allergies, and medications are clearly displayed on the profile page, but they will not be visible once the user navigates to other tabs. If the user forgets a patient's allergy, for example, the user could prescribe potentially lethal medication.

The patient's key information, allergies, and medications are persistently presented in the header, ensuring the information is always visible, regardless of which tab the user is on.

Presenting critical information

1 **Put critical information in the upper left corner of the screen.**

The Gutenberg Rule—which describes the general movement of the eyes when looking at a design—suggests the upper left corner of the screen is the primary optical area in cultures that read left to right, top to bottom.[3]

2 **Use color and text size to make critical information stand out.**

Gestalt theory—which aims to describe how humans perceive visual information—suggests that our attention will be drawn toward focal points created by contrast, such as color and text size. "Elements with a point of interest, emphasis, or difference will capture and hold the viewer's attention."[4]

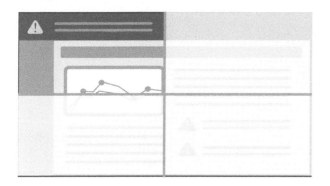

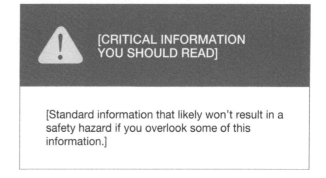

 Headers seem like a logical area to display critical information because they are prominent and persistent. However, a phenomenon known as banner blindness suggests that people tend to overlook information in banners and headers.[5] To help account for this, we recommend also including critical information within the screen's "body" (i.e., main portion).

Prevent users from disabling alarms

Principle
When an alarm informs users about a safety-critical condition that requires immediate action, the product should prevent users from disabling the alarm.

Why prevent users from disabling alarms?
Many products incorporate an alarm system to annunciate a dangerous situation requiring awareness and perhaps an immediate action. Alarms typically take a visual, audible, and/or tactile form. Usually more benign than an alarm, alerts might take similar forms. The effectiveness of alarms and alerts rides on them being enabled.

The consequence of turning off an alarm clock might not be harmful. But, turning off a patient monitor's heart arrhythmia alarm could be injurious or even fatal, failing to draw immediate medical attention to the ailing patient. Accordingly, certain alarms should be impossible to suspend or turn off, even if there is a potential for false and nuisance alarms. One reason to keep a critical alarm active is that the associated condition might require a rapid response. Another reason is that someone might forget to reactivate it.

A confident, or let's say overconfident, individual might be inclined to disable alarms, boldly believing that he or she is "in control" of the given situation. Unfortunately, such individuals might overestimate their ability to monitor situations vigilantly and, as a result, overlook a hazardous condition.

An exception to the guideline
In certain cases, it might be appropriate to suspend an alarm signal temporarily (i.e., to acknowledge it) to give the user a chance to respond to the condition without further distraction. These alarms should automatically reactivate within an appropriate period of time to continue offering protection.

Make alarms demanding

Ensuring that alarms remain enabled is good practice, but is not enough to ensure that alarms will be effective. Follow the tips below to increase the likelihood that enabled alarms will capture users' attention.

Use multiple communication channels

It helps to communicate alarms using multiple perceptual channels. Someone might be looking away from a visual alarm (e.g., flashing light) but still hear its aural counterpart (e.g., siren, beep, chirp). Someone who has a hearing impairment or is working in a busy environment might not hear an alarm signal but might notice a flashing light. Someone who is in a busy environment with a lot of visual distractions might not see or hear the alarm but might still be drawn by a tactile cue (e.g., vibrations such as those produced by most smartphones).

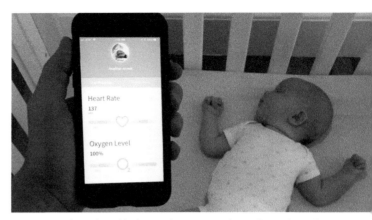

Many smartphone-based baby monitors communicate alarms through a combination of audible, visual, and/or tactile means.

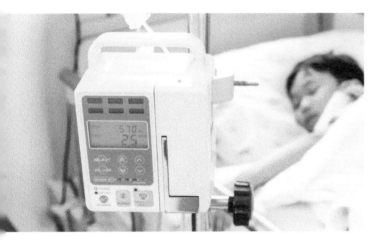

Many infusion pumps have a minimum alarm volume of 55 dB at 1 meter,[5] which increases the likelihood that alarms will be audible in a typical hospital environment, even when set to the lowest level. (Model photo: Colourbox.com)

Set appropriate alarm volumes and frequencies

Sometimes, lowering an audible alarm's volume is equivalent to disabling it because it will likely do a poor job of capturing a user's attention. As a general rule, users are likely to notice an audible alarm if it is 10 dB above the use environment's ambient noise level.[1] We also tend to best hear sounds that are in the frequency range of human speech: 1,000 Hz - 5,000 Hz.[2] Certain frequencies are also more attention-getting, such as those reached with a baby's cry.[3] Of course, sound designers should also keep in mind that we naturally lose our ability to hear high-pitched frequencies (those above 3,000 Hz) as we age (called presbycusis).[4]

Use optimal flashing frequencies

Flash rate influences a person's ability to notice a visual signal. This fact has led to the development of alarm system standards that prescribe optimal flash rates. For example, the National Fire Protection Association (NFPA) requires that fire alarms flash at a rate between 1 to 2 Hz (i.e., 1 to 2 times per second).[6] Advantageously, this flash rate was proven to be least likely to cause seizures in individuals.[7] Where multiple alarm systems might be visible at the same time, ensure they flash at the same rate.

Recent research has identified the phenomenon of alarm fatigue—a desensitization to alarms as a result of many frequent alarms. Some users, particularly in healthcare environments, resort to disabling alarms in an attempt to cope with alarm fatigue.[8]

Provide undo option

Principle
Give users an opportunity to undo an action to correct a recognized mistake, thereby avoiding or minimizing harmful consequences.

What if we could turn back time...?

We've all done it: accidentally clicked on a button we didn't intend to click. In such moments, we're grateful for the option of "undoing" our action and preventing any undesirable consequences. An "undo" option can give users the confidence to explore options and work efficiently, knowing that they can reverse or cancel actions. The absence of an undo option can cause users to feel "stuck," force users to perform cumbersome and/or dangerous work-arounds, or result in unrecoverable errors.

Consider a nurse who is trying to modify the infusion rate of a life-sustaining drug given to a critically ill patient. Now imagine that the nurse inadvertently set the rate to 102 mL/hr instead of 10.2 mL/hr. Without the option to quickly correct his mistake, he might have to restart the infusion pump, delaying life-sustaining therapy. Or, he might panic and remove the tubing set from the pump, which could result in "free flow" (i.e., unregulated flow) of drug to the patient.

So, the principle is a simple one: give users an option to "undo" wherever possible—especially in cases where not fixing a mistake might initiate a potentially dangerous action.

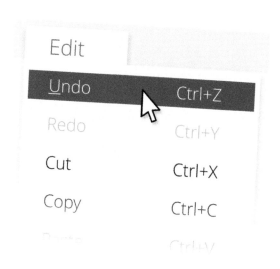

Let's go back to the beginning

We cannot be sure about when and where the concept of undo was first invented, but an early, documented example of an undo feature was in a computer-based word processing system developed at Brown University in the late 1960s.[1] Since then, undo features have been implemented in myriad types of software applications and products, ranging from medical record-keeping systems to online dating applications.[2] For most of us, use of the undo option, in the form of "Ctrl+Z" for Windows-based applications, "Command+Z" for Macintosh-based applications, or the ubiquitous "undo arrow," has become second nature.

Undo features can take many forms

There are many ways to enable users to undo a potentially dangerous action. Designers should carefully consider the scenario of use to determine which design is most likely to mitigate against potential use errors. Common undo implementations include:

Confirmation dialog
Enables users to confirm and, if desired, cancel an irreversible action before it begins, effectively acting as a preemptive undo feature.

Back button
In a linear workflow, enables users to return to a previous step or screen (e.g., to review, or revise previous selections).

Quit/Exit/Cancel
An extreme (and not always appropriate) solution, provides users with a way to abort a workflow or setting change.

Emergency stop
Enables users to quickly stop a process in a dangerous situation (see *Principle 51 - Enable emergency shutdown*).

Clear/Reset
Enables users to clear all entries and start from scratch, rather than clearing entries individually and risking leaving in old data.

Redo
As the opposite of "undo," enables users to repeat an action that they might have undone inadvertently.

Other design considerations

Below are some additional guidelines to ensure "undo" features are effective:

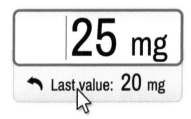

An "undo" function can present the previous value and provide a shortcut for reverting to the previous value.

- When enabling users to undo certain actions, such as entering or modifying data, provide an example of the resulting action. For example, present the original value that users would revert to by undoing the last action.

- Complement the "undo" label with a secondary label or description that explains which action will be undone (e.g., "Undo file deletion").

- For added safety, require users to confirm their intent to undo/cancel/quit with a confirmation dialog.

- In addition to software-based controls, include a physical button for urgent actions (e.g., STOP, Cancel) to help ensure users have immediate and direct access to the controls.

Avoid toggle ambiguity

Principle
Controls that toggle between two states should clearly indicate their current state.

What are toggle controls?

Toggle controls enable users to switch between two settings rapidly. Take the example of a light switch. Flip the switch up to turn the light on, and flip the switch down to turn it off. The switch's operation is intuitive not only because the popular convention holds that "up" is "on" and "down" is "off," but also because the resulting illumination or lack thereof provides immediate feedback regarding the switch's current setting. It's no wonder that the toggle light switch's design has not changed much since it was patented over 100 years ago in 1917.[1]

Another common type of toggle control is a single button that alternates between two states each time it is activated. For example, pressing the "Caps Lock" key on a keyboard causes all typed letters to be capitalized, while pressing the key a second time deactivates the feature.

Is this thing on?

Toggle controls that do not change position or do not provide immediate feedback can cause confusion due to a "toggle ambiguity" (sometimes referred to as "state-action ambiguity"). The ambiguity arises when a user cannot tell whether the toggle control's label describes the current state or the function to be performed. For example, a button labeled "On" might indicate that power is already on or that pressing the button will turn power on.

Toggle ambiguity can lead people to deactivate functions that they wish to activate and vice versa. Such use errors could lead to disaster in many cases, such as the operation of life-sustaining medical equipment (e.g., turning a blood pump on/off) or of an airplane (e.g., turning de-icing on/off).

Example: Power switch
Some users might assume that the switch indicates that power is "on" because the side of the switch with the line is toggled up. However, the switch is actually in the "off" position because the "O" is indented.

Example: Software-based toggle
Can you tell if the software-based switch is currently "on," or if it will turn "on" if you press the switch? It is unclear whether the "ON" label is the current state or the action.

How to avoid toggle ambiguity

There are multiple ways to prevent toggle ambiguity:

- **Label both states**
  Highlight the toggle control's current state, and indicate the state to which it will change if toggled.

- **Use verb-noun syntax**
  Replace labels that can be interpreted as either a state or an action with a verb-noun label that clarifies the control's function.

- **Explicitly indicate which option is selected**
  Present the selected option using a visually distinct color, and include an indicator (e.g., check mark, radio button) within the highlighted (i.e., active) option.

- **Use two separate controls**
  Replace a single toggle control with two controls that each represent a single action.

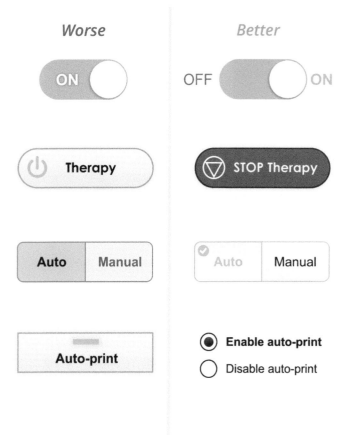

Protecting physical toggle controls

When toggle controls serve a critical function, they should be protected against inadvertent actuation. For example, if a toggle control used to activate a piece of machinery were unprotected and a factory worker bumped into the toggle switch, he or she could inadvertently move the switch from the "off" to the "on" position, activating the machinery and potentially injuring the worker. This type of use error could easily be prevented by recessing the control so it is not vulnerable to unintentional bumps.

Toggle switches have been used in automobiles for decades. Rows of toggle controls are often separated by physical barriers that help prevent the controls from being inadvertently actuated.

Some toggle switches used in aviation, auto racing, and other critical applications feature a physical guard that users must open in order to access the toggle switches.

Require professional maintenance and repair

Principle

Maintenance and repair of complex products, products requiring careful calibration, or products containing hazardous materials should be performed by professionals.

Professional knowledge is required to disassemble and reassemble complex products

Some products are so complicated and/or dangerous to disassemble that maintenance and repair tasks should be performed only by trained professionals.

In lieu of a professional's attention to it, a product might be damaged or reassembled—perhaps unknowingly—in a way that renders it inoperable or more likely to fail at a time when it is supposed to be serving a safety-critical function. Life support equipment falls into this category. That is why dialysis machines normally require factory-trained technicians to perform critical tasks, such as replacing a motor and recalibrating the machine.

Complex equipment requires both expertise and specialized tools to ensure correct and safe repair or maintenance.

Risks of disassembling products

What could be so dangerous about handling a product in a way intended only for professionals?

Exposing hazardous material

One potential hazard is that a product contains a hazardous material. An untrained individual could be exposed to the material and possibly release it into the environment.

Electric shock

Another hazard is electrical shock from internal components, such as capacitors, that retain energy for a long time, even when the device is turned off and disconnected from a power source. This can be the case with X-ray machines and televisions.[1]

Lacerations from sharp edges

A third potential hazard could be internal components, such as those with case parts with burred and sharp edges, that could cause lacerations and abrasions.

Minimize or prevent the need for maintenance

For the reasons already cited, some manufacturers engineer, advertise, and label their products in ways intended to prevent laypersons from attempting maintenance and repair tasks. Taken to the limits, they create "maintenance-free" products. This is the case with many 12v, lead acid automobile batteries that are permanently sealed.

It used to be that the owners of automobiles, tractors, and other kinds of machinery with a battery could check the electrolytic fluid level inside the cells. If the level was low, they could add more fluid. But, this action exposed people to the risk of getting battery acid on their hands, as well as the risk of battery explosion due to the presence of flammable hydrogen (a by-product of battery charging).

Some batteries still require refilling with acid, creating the potential for chemical burns.

Require special access

Another strategy to keep laypersons from breaking into a product intended for professional maintenance and repair is to require a special tool to open it—one that is unlikely to be found in a tinkerer's tool chest. In the case of a computer-based device, the tool might actually be a code that enables access to diagnostic software, as found in many new cars with computers that communicate electronically with so-called scan tools.

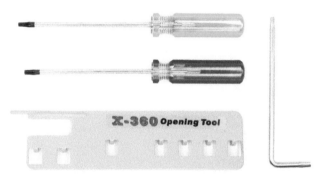

The Xbox 360 requires special tools to prevent laypersons from opening and tinkering with it.

Maintenance warnings

To combat the tendency among some individuals to undertake complex and dangerous maintenance tasks, some manufacturers include compelling warnings about the hazards and potential harms of doing so (see *Principle 79 - Add conspicuous warnings*).

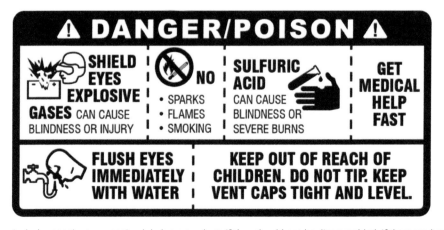

Include conspicuous warning labels on products if they should not be disassembled. If there are hazards associated with maintenance tasks, indicate the hazards, how to avoid harm, and the potential harm if the warning is not heeded.

Use telephone-style keypad layout

Principle

Products that require numeric entry should use a telephone-style keypad layout, which presents the 1-2-3- keys in the top row, to prevent incorrect entries due to muscle memory-related errors.

Different numeric keypad layouts

Safely completing a high-risk task, such as entering a medication dose on an intravenous infusion pump, can depend on entering the correct number. The wrong key press could lead to an injurious or fatal overdose or underdose. One culprit of such errors is a keypad layout that doesn't match the user's expectations.

Oh, I am not the type of phone you're talking about....

Numeric keypads on telephones and calculators usually have different key layouts. With little deviation based on the manufacturer, telephone keypads typically present the 1-2-3 keys in a row at the top, whereas calculator keypads typically present them in a row at the bottom. These solutions have rich histories. Human factors research at Bell Laboratories in the 1960s drove the decision to place the 1-2-3 keys at the top of new "touch-tone" phones. The research showed that a three-by-three matrix with zero at the bottom was easier to master than other options, including two rows of five numbers aligned vertically or horizontally, or numbers organized in circular configurations. One source posits that phone engineers chose the layout because they were concerned that people adept at using calculators featuring 7-8-9 along the top would enter numbers too quickly for the phone to recognize each button press. On the contrary, when Bell Labs asked calculator manufacturers about their decision to place the 7-8-9 keys at the top, they said the decision was arbitrary.[1]

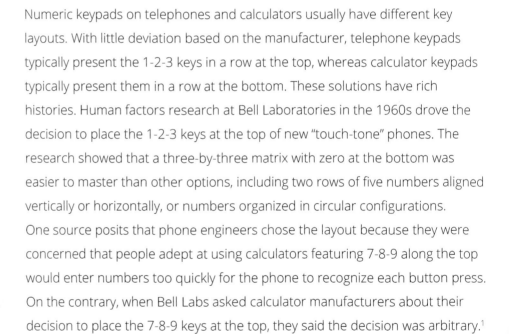

The original touch-tone phone had only a single key in the bottom row and did not have the letters "Q" or "Z."[2] The "*" and "#" keys were added in 1968.[3]

Preferred layout for all devices

The emerging consensus among human factors engineering specialists is that the telephone-style keypad is slightly better for applications requiring number pads.[4] The justification is that people frequently typing on telephones and using their mobile phones develop muscle memory of where to find the keys (sometimes doing so without looking). Muscle memory describes cases in which a person performs a task so many times that it does not require much conscious thought, thereby enabling the mind to focus on other things. Someone with strong muscle memory from using a telephone might make more errors with a calculator-style keypad (e.g., pressing "7" when they intended to press "1").

The NPSA recommends using a telephone-style layout for infusion devices because people are used to the layout from their mobile phones.

The National Patient Safety Agency's (NPSA) guidance document titled "A Guide to the Design of Electronic Infusion Devices" recommends using the telephone layout for all infusion devices.[5] Additionally, the NPSA recommends positioning the "0" and "." below the numbers and not next to each other, presumably to avoid inadvertently pressing the wrong key and changing the entered amount by an order of magnitude in either direction. That said, the NPSA recognizes that users looking at a telephone-style keypad, rather than at the screen, while entering numbers are prone to overlooking entry errors. The NPSA also recommends using analog methods like up/down buttons, which are more intuitive and require users to focus their attention on the screen when entering values.

The NPSA also recommends using up/down buttons because they require the user to look at the screen when entering values.

Some deviations...

Sometimes it is necessary to present the 0-9 keys in one or two rows to work within the constraints of a control panel's shape. But, this is something to avoid in cases where a conventional 3x4 array of sufficiently large keys will fit. Conceivably, data entry on irregular keypads is bound to be slower because the use of irregular keypads conflicts with muscle memory developed using conventional keypads.

Sometimes, a product's size and shape make it impossible to use a telephone-style layout. In these cases, one solution is to present the 0-9 keys in two rows.

Indicate unsaved changes

Principle

To prevent users from losing critical data, products should clearly indicate when critical data or settings changes are unsaved.

Safety first!

A common problem with software user interfaces is that they do not make it clear when changes to data will be lost if they are not saved. Such changes might include modifications to settings or entry field content. When the data is safety-related, uncertainty regarding the persistence of changes can spell peril. Consider an intensive care patient who is on a heart monitor. The monitor's default heart rate alarm thresholds might range from 50 to 120 beats per minute. Now, suppose a nurse wishes to narrow the alarm limits to 60 and 110 for a clinically indicated reason. The nurse might enter the new thresholds and assume they are "locked-in" without taking further action. But, what if the thresholds were adjusted in a pop-up window that required an additional confirmation, perhaps by pressing the "Save" button? And, assume that the nurse became distracted by another task and never pressed "Save." The thresholds could revert to their original settings and place the patient at elevated risk because she is not being monitored with the narrowed alarm thresholds.

Because the user never saved the changes, the alarm thresholds reverted back to the original settings.

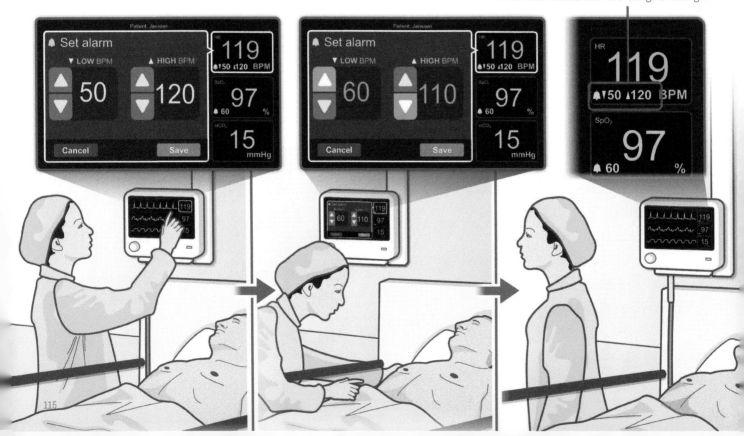

Don't let critical data go unsaved

The safe solution when dealing with critical data entry is to clearly indicate when changes take effect. The desired clarity might come from designing an application so that changes always take immediate effect (i.e., "What you see is what you get"—WYSIWYG). However, a WYSIWYG approach does not give users the opportunity to fix incorrect inputs before they take effect and potentially cause harm. Therefore, another common approach is to prompt users to confirm changes and then provide a message that the changes have taken effect. Similarly, the software could highlight unsaved changes, such as by showing a prompt next to an unsaved field. Whichever approach you take, apply it consistently throughout the product.

Vulnerability arises when the prompt to confirm changes automatically disappears without changes being saved. Accordingly, such messages should not automatically time-out. Or, if they must, the system should require users to acknowledge (e.g., dismiss) a message stating that pre-existing settings were restored. Similarly, if a user navigates away from a screen containing unsaved changes, the system should notify users of unsaved changes and prompt them to save or discard the changes.

Switching operational modes

Switching operational modes (e.g., from manual to automatic) should leave no doubt about whether the change was saved. That is why it is sometimes better to render the control in a hardware versus software form. In hardware form, the mode is always visually accessible, whereas a software-based control might be hidden by overlaying content or be found on a different screen altogether. One example of a physical mode change control is the control used on an Airbus A-320 aircraft to switch from "normal" to "ditch" mode. Ditch mode, which closes the aircraft's outflow valve and avionic ventilation ports, is critical for water landings like US Airways Flight 1549's landing on the Hudson River on January 15, 2009. Obviously, it is important for a pilot to know quickly and definitively whether the aircraft is in ditch mode. According to Federal Aviation Administration (FAA) regulations, all commercial airliners must have a ditch switch.[1]

Making it clear when changes take effect

What you see is what you get
With a "What you see is what you get" scheme, changes take immediate effect, eliminating the potential for unsaved changes.

Confirm changes
Many systems require users to confirm changes and will notify users if there are unsaved changes.

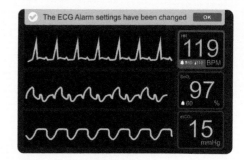

Provide feedback
A temporary message or banner can be used to provide clear feedback to users that settings were saved.

Make text legible

Principle

Text that conveys critical information should be legible to ensure users can read the text easily and accurately.

From your point of view

If you have ever had an eye exam, you know that text height is one of the most important factors influencing legibility. So is viewing distance. Together, these factors result in a subtended visual angle, as shown below:

← D (viewing distance, in inches) →

A

(Visual angle, in minutes)
(1 minute = 1/60 degree)

DANGER

H (height in inch[es])

Visual arc formula

In most cases, character height subtending a visual angle of **20 to 22** minutes assures legibility. An angle of **16 to 18** minutes might be acceptable, depending on several factors, including the text-to-background contrast ratio. Notably, **24 to 30** minutes is better for critical phrases and/or values (e.g., "DANGER").[1,2]

As an example, if you are standing 36 inches away, and the character height is 0.25 inches, you can use this formula to calculate the visual angle:

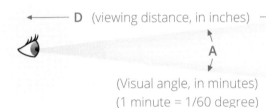

Angle = (3438* x Height) / Distance

24 min. = (3438 x 0.25 in.) / 36 in.

Or, you can rearrange the formula to calculate the character height based on a preferred visual angle:

Height = (Distance x Angle) / 3438*

0.25 in. = (36 in. x 24 min.) / 3438

For those who don't like math...

The US Federal Aviation Administration suggests using character heights of 1/200th the viewing distance to ensure legibility.[3] Using the previous example:

36 in. *(distance)* / **200** = 0.18 in. *(height)*

As a double-check, we can input 0.18 in. into our visual arc formula. At the same distance (36 in.), we find that it subtends 17.2 minutes of visual angle, which is within the 16 to 18 minute "acceptable" range.

To convert to font size

A "point" is equal to 1/72 inch. To convert a physical height to font size, use this formula:

0.25 in. *(height)* x **72** = 18 pt font

This is a convenient conversion. But, remember that point size includes the total height of a letter set (e.g., from the top of uppercase "F" to the bottom of lowercase "y"). Therefore, character height can vary depending on the word (e.g., "on" in lowercase vs. "ON" in uppercase).

*3438 can be used as a constant when the angle is less than 600 minutes. Otherwise, use the formula Angle = Tan-1(Height / Distance) X 60.

Contrast

To be legible, text must have sufficient contrast against its background. A ratio in the range of 3:1 to 7:1 is considered a minimum to ensure legibility,[4] depending on the application (e.g., smaller text should have a higher ratio). Therefore, seek to maximize contrast when information is critical; high-contrast pairings include black text on a light background or white text on a very dark background.

2:1	3:1	7:1
5:1	4:1	21:1

Examples of text-to-background contrast ratios[5]

Font style

One might think that Times New oman (serif) and **Helvetica** (sans serif) fonts, although different, are equally legible. However, in terms of maximizing safety, consider which you'd prefer on a road sign or warning label. Due to their simpler geometry, sans serif fonts are considered superior for rapid visual acquisition of shorter text strings (e.g., warning labels, on-screen prompts), whereas serif fonts are considered better at guiding the eye across long text strings.

More factors to consider

There are many other factors that affect legibility, such as stroke **thickness**, height / width ratio, l e t t e r spacing, and line spacing. Additionally, legibility can be impacted by user and environmental characteristics, such as:

- Visual impairments (e.g., retinopathy)
- How quickly the message must be read
- The stress level of a situation
- Lighting conditions

Poor legibility could leave important messages unread!

A user misreading or overlooking text due to its poor legibility can have serious consequences. Consider the following scenarios:

- A user perceives an instruction to be unimportant because the instruction is printed in small, barely legible text, leading the user to skip over the instruction. As a result, the user does not realize the device emits laser light and requires protective eyewear, consequently exposing himself to risk of eye injury because he did not don protective eyewear.

- A user misreads a label with poorly contrasting text, leading the user to pour windshield washer fluid into an automobile's power steering fluid reservoir instead of the windshield washer fluid reservoir.

- The poor legibility of text molded into the side of an automobile tire (due to the minimal contrast of light and shadow on 3-dimensional lettering) leads a user to misread the recommended pressure and inflate the tires to a dangerously high pressure.

- Small text leads a user to misread the recommended dose on a drug bottle label and, consequently, take too little or too much medication.

Provide backup display

Principle

If a display communicates critical information, provide a backup display to ensure that users have access to critical information in the event the primary display fails.

No display is fail-proof

Safety-critical displays must either be completely invulnerable to failures—often an unreasonable expectation—or must have a backup display. This principle certainly applies in high-stakes applications such as those often cited in this book: aircraft cockpits, nuclear power plant control rooms, and patient monitors. However, it applies to many more applications, such as the control systems found in ships, heavy lift cranes, and industrial factories.

A display may fail for various reasons and cause various degrees of information loss. In some situations, only part of the display might be rendered unreadable, whereas in other situations, the display might be completely out of service.

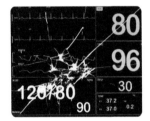

Physical damage

A cracked screen can reduce legibility or cause the display to fail altogether (see *Principle 7 - Temper the glass*).

Hardware failure

Hardware might fail suddenly due to electronic "gremlins" or ordinary events like being dropped.

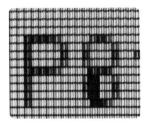

Dead pixels

Individual pixels can fail and, consequently, appear black (i.e., you have a dead pixel). Multiple dead pixels can affect legibility and cause reading errors.

Software failure

A physically intact screen can still be unusable if an underlying software failure results in corrupted data being sent to the display.

Physically separate backup displays

The security provided by redundant displays is greater when such displays are not vulnerable to common mode failures, such as a power surge in the same electrical circuit, a physical impact on the same portion of the associated device, or aging to the point of malfunction of the very same part. It is a positive sign when redundant displays are physically separated, differ significantly in design, and are electrically isolated.

Back up essential information

A backup display might merely duplicate the primary one or present supplemental information until it needs to serve in the primary role. Or, it might be a simpler, smaller, and lower-cost display intended to serve only in an emergency. Such a display might seem compromised in terms of its usability were it to serve in the primary role, but it might be just fine as a backup intended to convey critical information for a limited period of time.

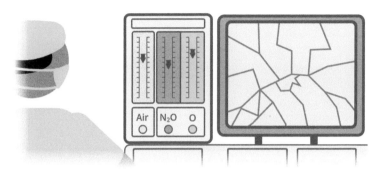

On an anesthesia workstation, physicians might normally check gas flow rates on a main display. However, if the main display fails, the physician could check the fresh gas flows by looking at a glass tube (i.e., a Thorpe tube) containing an indicator float.[1]

In certain aviation applications, an integrated standby instrument system serves as the backup to the primary flight display in so-called glass cockpits (i.e., those with an LCD display that replaces the traditional instruments). Pilots are encouraged to fly using the backup display on occasion so that they are better prepared to rely on it in case of a primary display failure.[2]

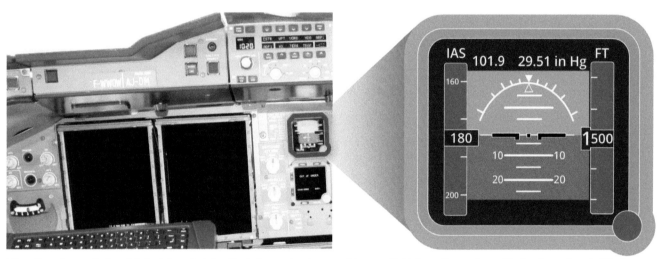

When the primary displays fail, the integrated standby instrument system provides essential information such as altitude, airspeed, and attitude.

Exceptions to the rule

One exception to the backup rule could be when a critical display failure is part of a larger-scale failure that renders the displayed information useless. For example, a failure mode that would damage a primary display might also render the entire device inoperable, in which case a backup display would serve no purpose.

Another type of exception is when a replacement device is readily available, and the time required to perform the swap is not critical. For example, if a hand-held glucose meter's display fails, the solution would be to buy another one. The risk posed by not being able to measure a blood glucose level at a given moment can be addressed by keeping an entire backup device on hand—an arguably practical solution when the cost is low and the device is reasonably small.

Use color-coding

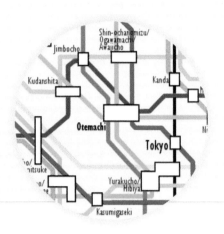

Principle

Use color-coding to distinguish critical information from non-critical information, draw users' attention to precautionary statements, highlight touch points, and help users identify and match components.

Design using meaningful colors

Color-coding is a powerful design tool. The human eye and brain (working as a team) are normally quite sensitive to color differences. Moreover, people tend to assign meaning to particular colors. Designers can utilize these traits to help ensure safe and effective interactions with all kinds of products.

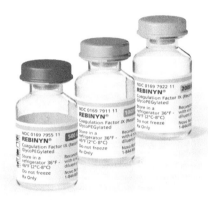

Many medication packages use color-coding to highlight differences in dosage or active ingredient.

An older, British-style train semaphore (i.e., visual signaling system) signals a train to stop. The choice for using red in train signals was primarily due to its clear appearance against different backgrounds (e.g., blue skies, gray clouds, green trees).[3]

Color codes are often abstractions, whereby the meaning of a color is learned rather than naturally related to a real-world characteristic. Red means "stop" and green means "go" in many cultures because people decided so. It could have been the other way around. According to one source, today's traffic light standard evolved from similar coding used in rail yards.[1] Another explanation for the use of red for "stop" is that red can be seen from the greatest distance, specifically because its wavelength is longer than that of other visible colors.[2]

In contrast, some color meanings are drawn from real-world characteristics. For example, it makes sense for a red light to indicate a hot stove burner because hot burners glow red. Blue is a good choice to communicate that something is "cold" because ice and cold lips can appear bluish.

Interestingly, different cultures associate colors with different feelings. For example, red can provoke alarm in the US, while it has a positive connotation in China.

Stand out from the crowd

Detecting a single visual item embedded among others can be difficult, even if there is some shape distinction. However, a difference in color can make an item "pop out." As such, the use of color can speed the task of finding the right control or information source from among many. Interestingly, the ability to find a uniquely colored element quickly does not depend on the number of other distracting elements surrounding it.[4]

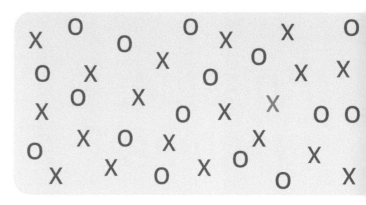

Color-coding precautionary statements

According to *ANSI Z535.4 - American National Standard for Safety Colors*, red, orange, and yellow should be used to communicate DANGER, WARNING, and CAUTION, differentiating hazards that (1) will cause death or serious injury, (2) could cause death or serious injury, and (3) could cause moderate or minor injury, respectively.[5]

RED = STOP

In many cultures, red means "stop," which is why emergency stop buttons are often red (see *Principle 51 - Enable emergency shutdown*). Imagine color-coding an escalator's emergency stop button ocher (brownish-yellow) instead of red (see image on the right). Not only would an ocher control be less conspicuous in most settings, but it would also be unlikely to communicate the message "stop" quite as effectively at first glance.

Use color-coding to guide users

Besides conveying meaning, color can be used to clarify that certain user interface elements are functionally related. In particular, color-coding can be used to create an association between on-screen elements and ensure that a cable or tube gets attached to the correct port. For example, it is fairly intuitive that a yellow-tipped cable should only be inserted into a port with a yellow ring around it (and not a port with a purple ring around it). That said, it would be even more intuitive if the matched items were also shape coded (i.e., keyed) to physically prevent a cable from being connected to the wrong port (see *Principle 82 - Add shape-coding*).

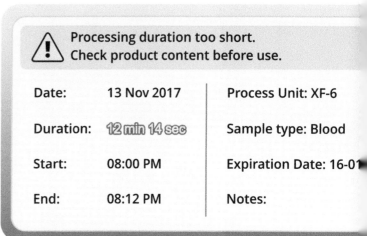

Users are likely to associate the yellow caution symbol with the underlying yellow text (and to do so quickly).

Exemplar 5
Anesthesia machine

Anesthesia machines are used by anesthesia providers to administer a mixture of anesthetic agents and other gases (e.g., oxygen, air, nitrous oxide) to patients through a mask or endotracheal tube. Over the past 100 years, anesthesia machine designs have evolved to include many safety features that have greatly improved patient safety.

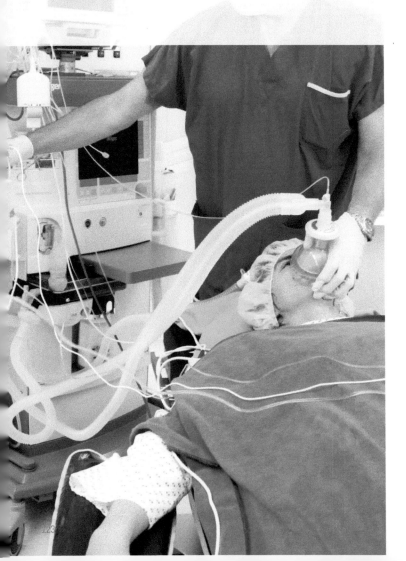

Prevent users from disabling alarms
Principle 42 - pg. 105

The anesthesia machine enables users to temporarily silence audible alarms, such as low oxygen pressure alarms, but prevents users from disabling an alarm altogether.

Display critical information continuously
Principle 41 - pg. 103

Continuous display of the heart rate, CO_2 and O_2 levels, airway pressure, and inspiratory and expiratory flow enables users to assess the patient's condition at any time.

Provide illumination
Principle 22 - pg. 61

Sensors recognize which gases are flowing and illuminate the associated controls and auxiliary ports.

Make text legible

Principle 48 - pg. 117

Critical information (e.g., vital signs) is displayed using large, high-contrast, alphanumeric characters to ensure users can easily read the information from the expected viewing distance.

Use color-coding

Principle 50 - pg. 121

Consistent colors identify the containers, controls, and monitor levels associated with N_2O (blue), air (yellow), and O_2 (green), enabling users to match them quickly and correctly.

Require professional maintenance and repair

Principle 45 - pg. 111

Trained technicians perform maintenance and repair tasks, which often require special knowledge and tools, to ensure the machine's safe operation.

Enable emergency shutdown

Principle
Enable users to shut down machinery if a hazardous situation is developing or has occurred.

Stop things when they go haywire

Some tools, machines, processes, and systems can quickly go haywire and pose a serious hazard. And, there are cases where a person might be in harm's way (e.g., a piece of clothing is caught in turning gears). That is why there are emergency shutdown controls (also known as E-stops). Such controls should be visually conspicuous, clearly labeled, large, and placed within the reach of users facing an emergency. Users should be able to actuate such controls quickly and with certainty that their action has triggered an immediate shutdown.

Motorcycles have "kill switches" in case the engine needs to be stopped immediately (see image below). For example, a motorcycle's rear wheel may keep spinning at high speed because the throttle is stuck in the open position after an accident. A driver, or passenger, pinned underneath the rear wheel can quickly turn the engine off to prevent (further) harm.

You also find emergency shutdown controls on somewhat simpler items such as table saws, which can bog down while cutting a piece of dense wood and need to be turned off immediately to protect the motor from overheating.

Emergency shutdown controls should be clearly visible and easy to access, and they should immediately interrupt the machine's operation. A motorcycle's kill switch can be operated while keeping both hands on the handle bar (left). Similarly, the emergency stop on an industrial scrubbing machine is easy to reach and is visually distinct and located away from other buttons (right).

How do E-stops differ from dead man's switches?

Although emergency stop controls and "dead man's switches" (see *Principle 23 - Add a "dead man's switch"*) have a similar function—to quickly disable a machine in a dangerous situation—they differ in how they achieve this endpoint. The dead man's switch requires operators to continuously actuate a switch to keep a system running. If the operator releases the switch (or becomes incapacitated), the dead man's switch automatically disengages and disables the machine. Emergency stop controls on the other hand, do not require continuous engagement. Instead, the user must actuate the emergency stop control to stop the machine.

Anatomy of an E-stop

There are multiple International Standards that specify the appearance of an emergency stop.[1]

Types of E-stops

There are three common types of E-stops, all of which remain in a "stop" position after being pressed:

Push-pull
The button is pushed in and locks into the "stop" position. It is released by pulling back.

Twist-release
The button is pushed in and locks into the "stop" position. It is released by twisting.

Key-release
The button is pushed in and locks into the "stop" position. The button can only be released with a key.

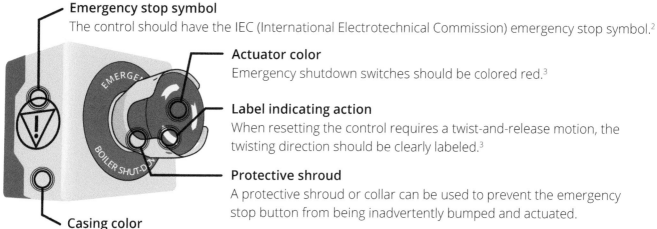

Emergency stop symbol
The control should have the IEC (International Electrotechnical Commission) emergency stop symbol.[2]

Actuator color
Emergency shutdown switches should be colored red.[3]

Label indicating action
When resetting the control requires a twist-and-release motion, the twisting direction should be clearly labeled.[3]

Protective shroud
A protective shroud or collar can be used to prevent the emergency stop button from being inadvertently bumped and actuated.

Casing color
The actuator's casing should be colored yellow.[3]

Types of machines that should incorporate an E-stop

According to EU Machinery Directive (2006/42/EC), electrical machinery in a wide variety of application areas must be fitted with one or more emergency stop controls:[1,4]

- Metalworking
- Wood production
- Textile production
- Food processing
- Compressors

- Lifting/moving equipment
- Robots
- Materials handling
- Electronic production equipment
- Inspection/testing equipment

- Printing
- Medical lasers & x-ray
- Packaging equipment
- Semi-fab equipment
- Pumping

Shield or isolate from heat

Principle
Protect users from burn injuries by shielding or isolating users from heat and by warning about thermal hazards.

Is it getting hot in here, or is it just me?

Every astronaut who has returned safely to earth owes his or her life to an ablative heat shield. This is obviously an extreme example, but the lesson is clear: it's important to protect people from contact burn injuries caused by things that get very hot—for example, a lithium-ion battery that overheats. Contact burns are typically classified as first-, second-, or third-degree burns depending on how deep they penetrate the skin's surface. Burn injuries can range from temporary discomfort to charring that can cause nerve damage.[1]

To protect users, heat shields should be solidly constructed to stay put during a product's service life, as opposed to tearing away at some point (i.e., due to wear and tear). Ideally, heat shields will remain relatively cool to the touch even when shielding an extremely hot component. However, any significant degree of protection is probably worthwhile even if the severity of thermal injury could only be reduced from a third- to first-degree burn, for example.

Why do we have reflexes?

When we touch a hot object, we normally have the reflex to recoil very quickly, typically within ¼ to ½ of a second.[2] However, if we do not recoil fast enough, we can suffer from contact burns. Burn exposure charts show how the skin typically responds to hot water at varying temperatures and exposure durations (see *Principle 80 - Prevent Scalding*). Such charts indicate that water heated to 148°F can cause a third-degree burn in just about 2 seconds.

First-degree
These types of burns include sunburns and cause skin redness that can be moderately painful.

Second-degree
These types of "partial thickness" burns penetrate the epidermis (the skin's top layer) and enter the underlying dermis, which can cause blisters and considerable pain.

Third-degree
These types of "full thickness" burns are the most severe and cause damage to the epidermis, dermis, and often the underlying tissue; they can result in charring, rather than blistering, and can destroy nerves.

Products that isolate users from heat

Over the years, many products have been improved to protect users from the dangers of heat. Some everyday examples include:

- **Heat-resistant toasters**
  Undoubtedly, the metal casings of vintage toasters caused many contact burns. Fortunately, most modern toasters are cool to the touch because their hot components are enclosed by an insulating material (e.g., heat-resistant plastic)—a smart move to protect the hands of curious young children and distracted adults.

- **Insulated handles**
  Sometimes, you need to use a protective towel to pick up a hot tea kettle. But, nowadays, many kettles are insulated or thermally-isolated for protection and comfort's sake. Similarly, modern frying pans have thermally-isolated handles that people can pick up without needing an oven mitt.

- **Isolated light bulbs**
  Traditional incandescent light bulbs can have an outside surface temperature in the range of 150 to more than 250°F.[3] Some contemporary light bulbs are relatively cool to the touch, largely because the light sources produce less heat to begin with (e.g., LED bulbs), but also because the bulbs have isolating ribs and/or shells that shield heat.

Worse   *Better*

Visual or audible heat indicators

When it is impossible to shield or isolate people from a thermal hazard, a product should provide some visual or audible indication that the product is or might become hot. For example, some stoves, especially those with a glass top, indicate when their surface is still hot by changing its color, displaying lights, or presenting an "H" (for Hot). These indicators warn an unsuspecting user or passerby to avoid resting a hand on the stove or placing a meltable item on the surface. Similarly, manufacturers might use warnings to communicate the nature of the hazard ("Hot surface"), the potential harm to the user ("Contact may cause burn"), and/or how to prevent the harm from occurring ("Allow time to cool before servicing") (see *Principle 79 - Add conspicuous warnings*). Audible indications, such as alarms, can also be used to notify users that a particular temperature has been reached or warn users that a surface is still hot and requires time to cool down.

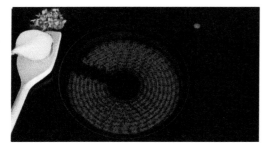

An electric cooktop burner glows red and uses a red LED to indicate that the surface is still hot and should not be touched.

A sample warning label warns users about a thermal hazard.

Install a physical shield

Principle

Install a physical shield to protect users from exposure to hazards such as heat, chemicals, radiation, flying debris, falling objects, or moving parts.

Shields are all around us (pun intended)

The word "shield" might conjure images of medieval knights protecting themselves from dragons, but today's shields are more likely to protect individuals from other types of flying objects, such as sparks and debris from a metal grinder. Nowadays, shields are also more varied and durable than those in King Arthur's day because contemporary materials and technologies offer protection against a wide range of hazards.

Shielding is necessary when people would otherwise be exposed to a harm in the normal course of using a device or machine, or in the event of a foreseeable failure. There are a seemingly infinite number of hazards that could cause harm if not shielded. For example, shields can protect against physical injury or damage that might be caused by fast-moving objects, and shields can also protect against spills, rotating blades, noxious gases, and radiation.

A shield can take many forms, such as a protective layer that the user dons or a physical barrier that is built into a product or workstation. Regardless of the application, it is important that one designs a shield that can protect a user without inhibiting the user's task. Various types of materials might be used for shields, ranging from plastic and thick glass to metal and chemical coatings. Other key factors that impact a shield's effectiveness include its location, shape, size, and strength (in part owing to its material).

Thin and flexible sunscreen
Sunscreen creates an almost imperceptible, protective layer from the sun's rays that enables a user to move freely and perform a number of activities unhindered.

2 mg/cm²
dosage[1]

Thick and durable bullet-resistant glass
Bullet-resistant glass's multi-layered construction shields the work area (e.g., an automobile cabin) against projectiles, while maintaining a user's visibility.

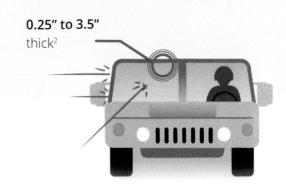

0.25" to 3.5"
thick[2]

In the absence of a proper shield, people could suffer harm in scenarios such as:

- Being contaminated by fluid spraying from a burst tube
- Being shocked by an energized electrical component
- Being burned by a hot heating element
- Being struck by flying debris from a rotating part failure
- Being exposed to radiation emitted from a medical scanner
- Being turned into lunch...

Given the variety of risks, such as those stated above, there are numerous ways to incorporate shielding. Consider which form of shielding offers the most effective safety mitigation for the specific use case. Samples of hazards and the associated shields to protect against such hazards include:

Chemical exposure
A negative pressure hood shields pharmacists from toxic chemicals and gases by venting dangerous gases and fumes away from the user and the surrounding environment[3] (see *Principle 27 - Eliminate or limit toxic fumes*).

Blunt force trauma
Sports equipment (e.g., helmet) fits snugly against specific parts of the body susceptible to harm and protects users from blunt force trauma and abrasions.

Flying debris
A transparent pane on a grinder protects workers from flying metal shards, while providing visual access to the work area.

Photo: Colourbox.com

Contamination
A suit with an integrated face shield protects surgeons from contaminated blood or other bodily fluids.

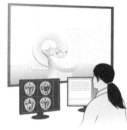

Radiation
A lead-lined window protects technicians from medical scanner radiation.

Falling objects
A hard hat protects construction workers from falling objects.

Spinning blades
A shroud (i.e., wire mesh) protects users' hands from contacting a fan's spinning blades.

Photo: Colourbox.com

Interference
Electromagnetic shielding protects users from electromagnetic energy that escapes from or penetrates electronics.[4]

Make blades very sharp

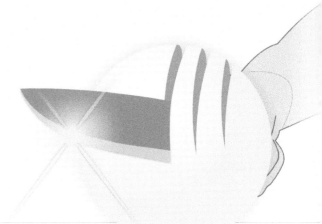

Principle
Make blades very sharp to reduce the force required to produce a clean cut and give the user tight control over the blade.

Sharper means safer

It might sound counterintuitive, but a dull blade can be more dangerous than a sharp one.[1] If a product is intended to cut through material, it's best if it does so with ease. This applies to scalpels used in surgery, blades spinning within table saws, kitchen knives cutting vegetables, and the blades within razors used for shaving.

A dull blade is an obstacle to the task at hand—to cut something. It might take more force to chop or slice through material when using a dull blade. Therefore, you have a task involving gross motor motions, rather than fine motor motions. As a result, some control is lost. A dull blade might also skip along a surface, such as a tomato's skin, rather than cut into it. This unwanted motion could put the user and others at risk of being cut. The same principle applies to chisels used by woodworkers. A very sharp chisel makes it far easier to control the tool and requires less force to produce the cut. Conversely, a dull chisel can hinder wood cutting, induce hand strain, and cause wayward chisel movements that could cause the blade to strike the user.

Dull blades require users to apply more force. A moment of inattention could result in a nasty cut as the blade slides over, instead of into, a piece of fruit.

"Scalpel, please..."
In surgery, a sharp blade helps ensure a clean, rather than ragged, cut. In turn, a clean cut enables more effective suturing and wound healing.[2] To ensure their blades are as sharp as possible, many surgeons use new blades with every patient, albeit often with a reusable scalpel handle. The alternative is to use an entirely disposable scalpel like the one pictured on the right, which typically has a plastic handle and a retractable blade (similar to a utility knife).

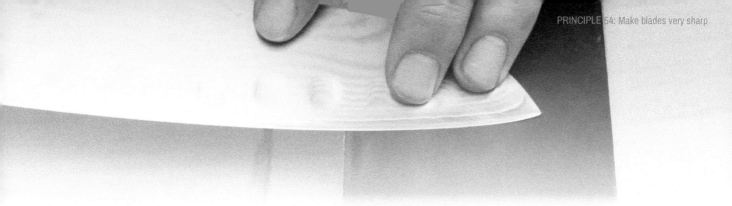

Keeping knives sharp

It's quite convenient when products incorporating blades can sharpen themselves or have holders that do the trick. One particular manufacturer offers knives that are stored in a knife block containing slots that are engineered to sharpen a knife with every withdrawal and reinsertion.[3] One chainsaw has a built-in mechanism that simply requires the user to pull a lever to expose the moving chain to an internally mounted sharpener.[4] Chain-sharpening attachments are also available to ease the task and help ensure the tool cuts safely and effectively.

Of course, there is the tried-and-true method of blade sharpening using muscle and a sharpening device. There are plenty of standalone sharpeners that simply require one to draw the blade through a groove. There are also a wide variety of grinders, sharpening stones, and devices with an equivalent purpose that give the user the pleasure of gauging the blade angle and doing his or her own shaping and honing.

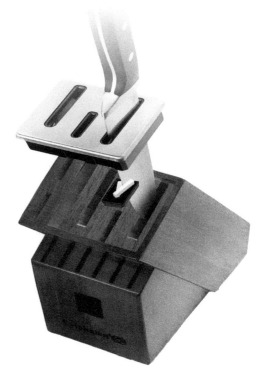

This knife block sharpens blades each time the user removes and inserts them, ensuring the blades remain sharp.

Durability and sharpness

Most blades are made from some sort of steel, with stainless steel and carbon steel serving as particularly good choices to produce a very sharp edge. Ceramic blades can be incredibly sharp and a great choice in the kitchen as well.[5] One benefit of ceramic blades is that they never require sharpening!

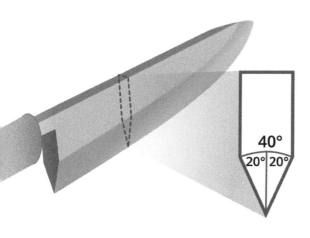

Sharpness increases with the edge's bevel angle. A bevel angle of 15-20 degrees (measured on one side relative to the blade's central axis) is enough to produce a sharp edge. A knife sharpened to 20 degrees on each side (i.e., double beveled) will have beveled sides that form a 40-degree angle. When the angle gets smaller, sharpness increases, but so does the tendency for the thinner blade edge to deform and become blunt with use, thereby requiring more frequent sharpening.[6]

Start on slow and low

Principle

Products for which a high setting could be dangerous should start up slow (i.e., at a low power level) and then require deliberate user action to switch to a higher power level.

Don't get blown away!

Products that can cause sensory and physical strain should give people a chance to transition to the higher state of engagement—to get up to speed, so to speak. Consider a car stereo capable of emitting a very high volume. Clever ones will power on at a low volume even if they were originally turned off at a high-volume setting. This prevents the shock of blasting tunes. More importantly, it protects sensitive eardrums, at least until someone chooses to crank it up to a potentially damaging volume level (see *Principle 4 - Limit sound volume*). A rotary encoder that does not have fixed setting points can be used in any product where full power is not required immediately, and full power could lead to harm. Notably, such an encoder would be best served by a digital readout.

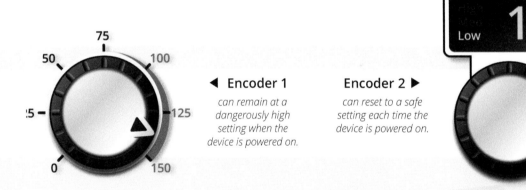

◀ **Encoder 1**
can remain at a dangerously high setting when the device is powered on.

Encoder 2 ▶
can reset to a safe setting each time the device is powered on.

Walk before you run

Treadmills are a good example of a product that starts slow to prevent accidents. Exercisers must actively command their calorie-burning machines to accelerate because starting at a high speed could cause injurious falls. Notably, the same basic principle applies to the low engine speed applied when placing a vehicle into first gear (or Drive) and requiring the user to engage a spring-loaded throttle (e.g., foot pedal) to increase speed.

No!

Start slow and low

Personal differences

The principle of starting "slow and low" can also help adapt products to users' personal differences, which might vary greatly across user populations or over the course of a given user's experience with the product. For example, many healthcare products and therapies are designed to start at conservative settings/doses to minimize the potential for discomfort or adverse effects, while enabling users to modify the settings/doses as needed to accommodate personal preferences.

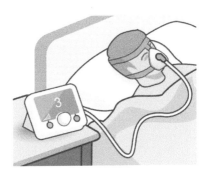

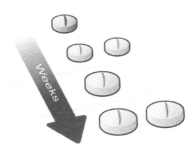

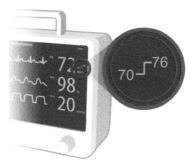

Personal comfort

Some continuous positive airway pressure (CPAP) devices, which treat sleep apnea, slowly increase airway pressure over time so that new users become accustomed to the sensation of air moving into their windpipes.[1] Without this feature, people might abandon the therapy due to the initial discomfort.

Physiological effects

To give a person time to adjust to new medication and monitor the medication's effects, some medication regimens start with a low dose and ramp up to a target dose over several weeks.[2] Similarly, when ending a medication regimen, it is common to reduce the dose slowly over several weeks, rather than stopping immediately.[3]

Monitoring preferences

Many patient monitors feature "narrow" alarm limits by default, which make the monitors more sensitive to potentially dangerous changes in patient conditions. However, the monitors enable users to broaden the alarm limits to accommodate certain patient populations or medical conditions and decrease the potential for false alarms.

Learning to fly

In a sense, training is another form of starting slow and low. Many products require training, and in some cases, users continually progress toward using more complex and powerful features or even versions of such products. For example, new pilots learn to fly low power, single-engine propeller airplanes before progressing to more powerful, multi-engine airplanes, and then jets.[4] This gives pilots the chance to gain experience with planes that are less complex before attempting to fly those that require more skill.

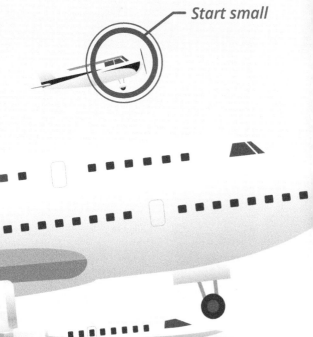

Start small

Use sensors

Principle
Use sensors to continuously monitor and detect events or changes in the environment that could put users at risk.

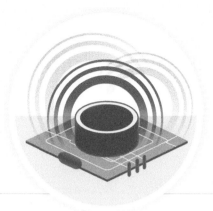

Sensors augment us and our devices

Sensor technologies extend our ability to attain critical information that we might otherwise ignore or overlook. In the same vein, sensors can be thought of as the first component of a predictive system (see *Principle 18 - Predict hazardous situations*), which might foretell a complex scenario with safety-critical implications. Many deaths and injuries could potentially have been prevented through the use of various sensing technologies, such as those listed below:

Baby monitors can detect a baby's cry while filtering out other noises, thereby reducing false alarms and alerting parents to a baby in distress. There are also more advanced monitors that include a smartphone app and a wearable sensor that can measure an infant's oxygen level and temperature.[1]

To prevent a drunk driver from operating motor vehicles, law makers and car makers are considering equipping cars with sensors that detect alcohol in the vehicle's air or on the driver's fingertips.[2,3]

Some table saws sense when a finger touches or approaches a spinning blade and stop the blade's movement before the user can even react, typically leading only to a small scratch or cut, rather than a severe injury to the finger.[4]

Smoke, carbon monoxide, and explosive gas detectors can detect hazardous situations, prompting users to react or evacuate before it's too late. One study showed that the death rate per 100 reported home fires (in the US) was more than twice as high in homes without a working smoke alarm.[5]

Function allocation: user vs. machine

Although sensors can perform certain tasks better than humans, we excel at certain tasks that might challenge sensor-based technologies. To help identify optimal uses for sensors, designers can conduct a "function allocation analysis," which helps compare the strengths of humans versus machines at specific tasks.[6] Designers can use the results of this analysis to design systems with sensors that can detect safety issues that might fall outside a human's ability to reliably sense.

Humans are good at:	Machines are good at:
Uncovering creative solutions	Vigilant monitoring
Dealing with change and improvisation	Performing repetitive tasks correctly
Visual information processing	Applying logic
Inductive reasoning	Performing at high speed

Some of the functional strengths of humans and machines.[7]

Examples where devices compensate for our limitations include:[7]

Processing complex data
Machines can quickly analyze complex data and synthesize the results into actionable information.

Providing constant vigilance
Machines can monitor consistently and detect subtle signals with a high level of accuracy, whereas users can become distracted or fatigued.

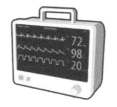

Confirming critical interactions
Systems can detect correct assembly and alert the user to incorrectly installed or missing components, rather than relying on users' visual inspections.

Accelerating detection
Devices can react quickly and at precise thresholds that might exceed a user's ability to detect and react.

Sensor-y overload!

With the advent of myriad Internet-connected smart devices, there is an inherent risk that users become overwhelmed with unnecessary (i.e., non-actionable) and/or difficult-to-decipher data, simply because it is available. This overload could lead to anything from minor distraction to potentially severe outcomes, such as missing critical alarms due to alarm fatigue.[8]

One cloud computing expert proposed questions we can ask before deciding to make a device Internet-connected:[9]

"Does this solve an existing problem, and can humans benefit? Are there manual alternatives that are just as effective? Is the use of the technology cost-effective? Is data being consumed that will provide us with a deeper understanding of the activity?"

Enable escape

Principle
When a product can encapsulate a user, provide
a means of quick escape.

Escaping public transportation

In some cases, maximizing safety might require enabling passengers to rapidly exit from a vehicle during an
emergency. Mass transit vehicles and aircraft are good examples of such vehicles, where large groups of people
could become trapped during an accident and need a way to escape the wreckage. As such, easily accessible
escape mechanisms are needed, in part due to the number and variety of users in such vehicles and the fact
that passengers might be unfamiliar with the mechanical functions of the vehicle they are in.

Airplanes have many features that guide people to exits,
including safety videos and safety cards that indicate
exit locations, floor lighting that highlights pathways
to exits, directional signs, and door operation
labels. One or more of these features might help
passengers find their way out of an airplane
in the event of an emergency evacuation.
Notably, the Code of Federal Regulations
(14 CFR) part 25 calls upon manufacturers
of planes with more than 44 seats
to demonstrate that a full load of
passengers can evacuate in 90
seconds or less using half the exits.[1]

*The Airbus A320 has six different exit locations to enable
passengers to evacuate quickly in an emergency.[2]*

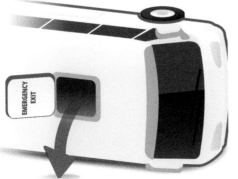

Trains and buses label windows that can pivot open when normal exits
are rendered unusable. Some trains and buses have a hammer or
button attached to the window, which enables passengers to break the
tempered glass and exit. Some buses also have a hatch in the ceiling to
enable escape if the front door is inoperable or if the bus is submerged
in water.

Hotels and public venues

Movie theaters, shopping malls, and restaurants feature exit signage and emergency exit doors. In many countries, hotel rooms are required to include a map showing the hotel's exits and escape routes. These maps often include other potentially helpful information, such as the location of fire extinguishers and stairs, in case the elevators become unsafe to use.

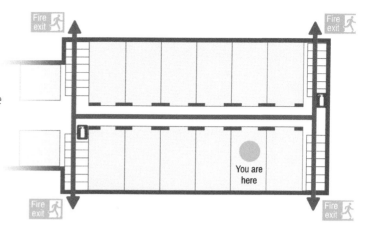

Escaping your personal possessions

Our homes and daily lives include multiple examples of products and scenarios that should allow for quick escape to ensure safety.

Today, refrigerators are designed to enable an entrapped child (or anyone for that matter) to push the door open from the inside. By comparison, older models could latch closed, causing tragedies involving children playing inside of discarded refrigerators.[3] Similarly, kids' toy chests should never lock or even latch closed in a way that would prevent an entrapped child from escaping.

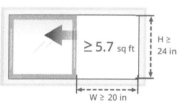

To enable occupants to escape or emergency response teams to enter, international residential building codes call for all bedrooms to include an emergency exit to the outside. This could be a door, but it typically takes the form of a window, which must meet minimum requirements for height, width, and overall area (among other requirements).[4]

In 2001, the US National Highway and Traffic Safety Administration (NHTSA) responded to accidents involving children getting stuck in automobile trunks by requiring new passenger cars (starting with model year 2002) to include a glow-in-the-dark trunk release lever, which enables trapped passengers to open the trunk from inside.[5] In 2012, the safety watchdog and advocacy group Kids and Cars reported that, "since the mandate, no children have died in the trunk of a car equipped with a release lever, but accidental trunk entrapment in older cars has killed at least 22 children since 2002."[6]

Many ways to enable escape

Escape features do not always take the form of hatches, doors, or well-planned evacuation routes. Rather, escape features might be accessories, such as lifebuoys, elevator emergency phones, gas or oxygen masks, and fire extinguishers, which all enable users to escape dangerous situation or call for help. In any case, make sure escape-related instructions are easy to find and see, and that escape mechanisms are accessible, usable, and function as expected in the use environment.

Detect fatigue and rouse users

Principle

Design systems so they can detect when users are drowsy or asleep, and then rouse users to prevent accidents due to fatigue or drowsiness.

Operating vehicles + fatigue = danger

Falling asleep during high school math class has consequences, but ones that are not as dangerous as the sort that can occur when people fall asleep, or even become drowsy, while performing critical tasks.

The US National Highway Traffic Safety Administration (NHTSA) reported that in 2014, NHTSA's FARS database contained 846 documented fatalities related to drowsy-driving (accounting for 2.6% of all fatalities). And, the problem is not a new one—from 2005 to 2009, there was an estimated average of 83,000 drowsy-driving-related crashes each year.[1]

Is this a worldwide problem? Of course! People are largely the same when it comes to how they perform when they're fatigued. In fact, over 2,000,000 accidents and approximately 190,000 fatalities per year worldwide are due to driver fatigue.[2]

Pilot and air traffic controller fatigue is also a recognized problem. This is reflected in US Federal Aviation Administration (FAA) rules established in 2011, which limit working hours for pilots. Now, on-duty time is limited to 9-14 hours depending on the flight route and when the pilot's day begins. Pilots must also have a rest period of 10 hours (with the opportunity to sleep uninterrupted for at least eight hours) before returning to duty.[3]

Fatigue and mental effectiveness in surgeons

There is also cause for concern regarding sleepy surgeons. As reported in one study, 27 physicians in training (i.e., residents) got approximately five hours of sleep per night on average. An analysis of their mental capacity was disconcerting. The results showed that 48% of the time, residents functioned at less than 80% of their full mental effectiveness due to fatigue. And, 27% of the time, they functioned at less than 70% mental effectiveness.[4]

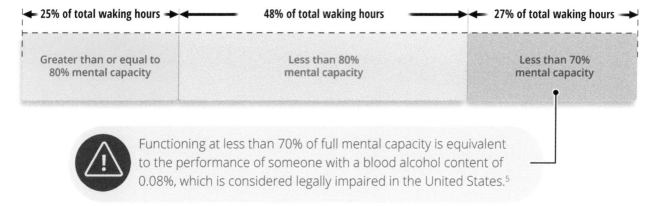

◄ 25% of total waking hours ►	48% of total waking hours ►	◄ 27% of total waking hours ►
Greater than or equal to 80% mental capacity	Less than 80% mental capacity	Less than 70% mental capacity

⚠ Functioning at less than 70% of full mental capacity is equivalent to the performance of someone with a blood alcohol content of 0.08%, which is considered legally impaired in the United States.[5]

Detecting drowsy or sleepy drivers

Automobile companies (e.g., Audi, Mercedes, Volvo) and construction machinery manufacturers (e.g., Caterpillar) have tried to address the hazard of fatigued driving by developing systems that detect and rouse drowsy and asleep drivers.[6,7] The systems detect behavior such as the following:[8]

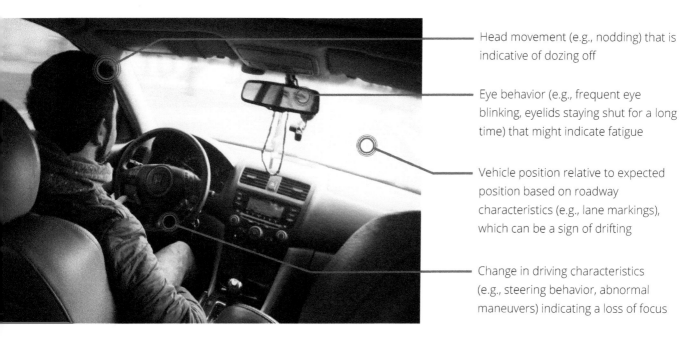

Head movement (e.g., nodding) that is indicative of dozing off

Eye behavior (e.g., frequent eye blinking, eyelids staying shut for a long time) that might indicate fatigue

Vehicle position relative to expected position based on roadway characteristics (e.g., lane markings), which can be a sign of drifting

Change in driving characteristics (e.g., steering behavior, abnormal maneuvers) indicating a loss of focus

How does an alertness-monitoring system perk up a driver? Various systems use different methods, including visual alerts, audible alerts, and vibrating steering wheels. Visual alerts have been presented on dashboard displays and on heads-up displays. Audible alerts have included tones and synthesized rumble strip sounds. One study from 2006 suggested that vibration was a particularly effective way to rouse the drowsy driver.[9] Similarly, shaking the control yoke (in other words, the pilot's steering wheel) works to alert pilots that their plane is about to stall (see *Principle 8 – Provide tactile feedback*).

Augment control

Principle
Design products, systems, and machines to take over or supplement human control in cases when the need for control surpasses humans' abilities or enables safer operation.

Driving toward greater control

Have you ever been driving on an icy winter night, when you slow for a stop sign ahead and feel your car's brake pedal pumping under your foot? This system, known as the anti-lock brake system (ABS), is perhaps the most ubiquitous example of enhanced control. Lacking ABS, an automobile is more likely to skid during hard braking. Drivers used to be advised to "pump the brakes" to balance braking performance and steering effectiveness. Now, drivers can simply step on and hold down the brake. This lets the system do the braking in an effective manner that prevents skidding, while enabling the driver to steer to avoid hazards.

Having proved its worth, ABS was required in new automobiles in the European Union countries starting in 2004.[1] Although the United States had mandated ABS on trucks and buses in 1999,[2] the US government was slow to adopt a mandate requiring ABS in cars. Finally, in 2011, cars sold in the US were required to incorporate an Electronic Stability Control (ESC) system, which uses automatic computer-controlled braking.[3] Now, some cars can even detect hazards (e.g., an object in the road) and automatically apply the brakes to avoid a collision (see *Principle 18 – Predict hazardous situations*).

Augmented control in aviation
Another impressive example of augmented control is the fly-by-wire (FBW) system installed in many aircrafts. Systems such as the one found in some Airbus airliners compare a pilot's inputs to the plane's aerodynamic capabilities to optimize control. For example, a pilot can command an Airbus 320 to climb rapidly, but the flight computer will enable the plane to climb only as fast as it is able to avoid a dangerous stall, providing "flight envelope protection." Despite the safety benefits of such systems, aviation experts continue to debate the wisdom of taking ultimate control away from pilots,[4] noting the important ability of humans to adapt to novel circumstances.

Smart excavating equipment? We dig it.

The next example of augmented control is the backhoe, which is a type of excavating equipment. New models can be configured to automate certain operations. For example, a backhoe can be set so the bucket digs to a specific depth and no deeper. This can help avoid underground hazards (e.g., buried gas and powerlines) as well as speed up digging operations.

Let's raise this bucket up!

Sorry, I can't go any higher—powerline overhead!

Similarly, a backhoe performing roadwork can be set up so that the bucket does not swing past a boundary and into roadway traffic or an overhead powerline, for example.[5] Of course, operators must remain vigilant when it comes to checking bucket swing pathways for hazards and should not rely solely on automated movements. But, the augmented control can be just the thing needed to avoid common accidents.

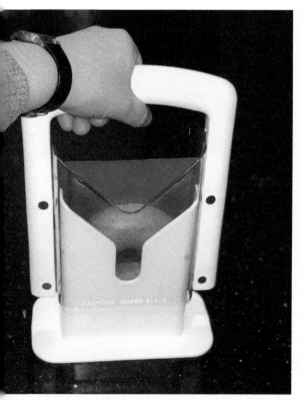

A simple kitchen tool, such as a bagel cutter, helps reduce finger injuries due to bagel slicing.

"I'll have an everything bagel—sliced please"

In addition to electrical or system-based augmented control, there are purely physical ways of augmenting control to enhance safety. Take for example the task of slicing a bagel in half. Something as simple as a slotted wood block with flexible sides offers protection against cuts by holding a bagel tightly during slicing. Personal safety may be extended further by taking the knife out of the mix and using a bagel cutter that simply requires the user to push the bagel through a slicer— the solution used in my restaurants.

Did you know that in 2008, Americans ate an estimated 3 billion bagels at home? Did you also know that trying to slice a bagel landed 1,979 people in the emergency room with cut fingers? According to Freakonomics.com, this fact makes bagel-cutting the 5th most dangerous activity in the American kitchen![6]

Reduce (or isolate from) vibration

Principle
Products should eliminate, isolate, or reduce exposure to vibration that could cause vibration-related injuries, such as hand-arm vibration syndrome (HAVS).

Shaking hands is not always a good thing
Certain amounts of vibration can cause ill health effects. Such injuries are cumulative and related to the vibration's intensity, duration, and frequency of exposure. For example, limited exposure to an intensely vibrating jack hammer is less likely to cause injury than long-term exposure to a less intensely vibrating drill.

Hand-held tools that vibrate intensely can eventually injure arms, hands, or fingers. The resulting condition, called hand-arm vibration syndrome (HAVS), can involve neurological, vascular, or musculoskeletal damage. Symptoms include tingling and numbness, pain, reduced strength and range of motion (i.e., mobility), and a whitening of the fingertips due to lack of blood circulation (i.e., a secondary Raynaud phenomenon).[1]

Beyond the hands and arms, vibration applied to the whole body has been linked to problems such as back pain, digestive issues, prostate issues, and miscarriages.[2]

If a product must vibrate to perform its function, seek ways to dampen vibrations and reduce the amount of time the product is used.

 The National Institute for Occupational Safety and Health (NIOSH) and the Directive 2002/44/EC from the European Agency for Safety and Health at Work call upon employers to consider the vibration-related risks to workers and to provide appropriate protections. Additional quantitative guidance can be found in ISO 2631-1 (1997), ISO 5349-1(2001), and 5349-2(2001).[3,4]

Image above simulates the possible appearance of fingers affected by white-tip syndrome.

Vibrations of 0.1 to ~1,500 Hz are the most relevant to human work measures[5]

0.1 - 0.5 Hz
Motion sickness[6]

Those who suffer from motion sickness can attest to how debilitating low-frequency vibration can be—for example, the rocking of a boat or the turbulence of an airplane. Taking certain medications or simply staring at the horizon might alleviate symptoms. However, perhaps the best approach is to steer clear of turbulent environments.

1 - 80 Hz
Whole body vibration[5]

Truck drivers can be injured by the constant vibrations produced by their trucks' engines and suspensions. Some truck seats use passive means, such as springs and padding to minimize vibrations, while other seats use active dampening (e.g., a system that detects bumps and uses vibration-canceling technology to dampen them) to reduce vibration transmission.[7]

8 Hz - 1,000 Hz
Hand-arm vibration[5]

Vibration injuries to the hands can be caused by large tools, such as jack hammers, but also by small tools, such as dental drills. Vibration-reducing gloves can reduce vibrations transmitted to the palm by 5%–58% (depending on the tool).[8] Tools can also be constructed of metals (e.g., tungsten alloys) that dampen vibrations[9] or can include an auto-balancer to offset imbalances from rotating parts.[10]

Avoiding exposure

In addition to designing vibration-reducing features, proper technique can also play a role when making interactions with vibrating products safer and more comfortable. For example, gripping tools less firmly and letting them "do the work," rather than "forcing them," can lessen the vibration transmitted to the user. Another safety tip is to avoid using vibrating tools with cold, damp hands (coldness can reduce circulation to extremities).[3] More generally, examine tasks as a whole to find any opportunities for reducing exposure time. For example, replacing a cutting tool's dull blade with a sharper one could reduce cutting time, thereby reducing exposure to the vibrating tool (see *Principle 54 - Make blades very sharp*). A more effective exposure avoidance technique would be to mount a device on an isolated stand whereby users no longer need to hold the vibrating device at all.

Workers can reduce vibration transmission by gripping tools less firmly and by wearing gloves that keep hands warm and absorb vibration. (Model photo: Colourbox.com)

Exemplar 6
Chainsaw

Chainsaws are inherently dangerous. They are hand-held cutting machines with engines that can produce several horsepower. And yet, modern chainsaws have an impressive set of safety features that greatly reduce risk.

Add a "dead man's switch"

Principle 23 - pg. 63

Users must hold down the throttle trigger interlock continuously to operate the chainsaw. If users release the trigger, the chain stops.

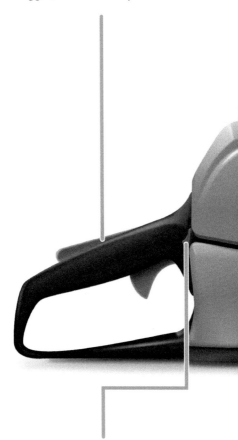

Enable emergency shutdown

Principle 51 - pg. 125

Stop switch located next to the operator's thumb enables the operator to shut down the engine immediately to avoid a dangerous situation.

Make it slip resistant

Principle 40 - pg. 99

Textured, elastomeric handles provide a secure grip that helps users control the chainsaw while minimizing arm and wrist strain.

Install a physical shield

Principle 53 - pg. 129

Shield protects the operator's front hand from the blade and flying debris.

Make blades very sharp

Principle 54 - pg. 131

Sharp teeth make cutting easier and reduce the chance of kickback, which can occur when the chainsaw's blades become caught in the object they're cutting.

Reduce (or isolate from) vibration

Principle 60 - pg. 143

Integrated vibration dampeners minimize the negative effects of prolonged exposure to vibration, such as numbness and reduced strength.

Enable fast action in an emergency / Use sensors

Principle 90 - pg. 209 / Principle 56 - pg. 135

Chain brake system stops the chain in less than a second when the chainsaw rapidly "kicks back" toward the user or when the operator activates it manually.

Shield or isolate from heat

Principle 52 - pg. 127

Housing shields operators from heat generated by the chainsaw's engine, while the muffler guides hot exhaust away from the user.

Prevent fluid ingress

Principle
Products used around fluids or in wet environments should be designed to prevent water ingress, ensuring they remain operable and safe.

A little splash never hurt anyone, right?

The precautions on the right are good and necessary advice if a device is not waterproof and subject to failure if it or its internal components become wet.

An electrical device might pose a shock hazard if it gets wet, or at least short-circuit and become inoperative, which by itself could pose a hazard. For example, what good would a marine radar unit be in a storm if it stopped functioning after being splashed? A device containing dry raw materials, such as a powdered (lyophilized) drug, also might not be safe or efficacious to use if soaked.

The obvious solution is waterproofing, or at least designing a product to be relatively invulnerable to fluid ingress. There are plenty of standards for what constitutes being waterproof and water resistant.

Gaskets, O-rings, glue, and caulking are common means to waterproof items comprised of joined parts. Another solution is to place a vulnerable item inside of a waterproof container that is waterproofed by the aforementioned means.

 Never immerse device in water!

 Do not expose to moisture!

 Avoid splashing fluid onto device!

 Keep dry at all times!

Gaskets create a mechanical seal by filling the space between mating surfaces. For example, a gasket integrated into a digital camera's battery door creates a seal between the battery door and the camera body, preventing water and particulates (e.g., sand, dirt) from entering the battery chamber.

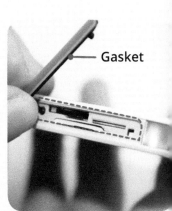

Gasket

O-rings are one of the most common gasket types, dating back to the late 1800s. These simple gaskets consist of a donut-shaped ring with a circular cross section. They are widely used because they are inexpensive, reliable, and can maintain a seal even under high pressure.

Getting deeper

Water-resistant products might use mechanical means to protect against downpours or splashes, such as those created by Mother Nature (i.e., rain) or people who spill their soft drink. For example, there are deflectors, collection gutters, and covers.

Products face an increasing waterproofing challenge as water pressure increases exponentially with depth. Accordingly, manufacturers often claim waterproofness only to a specified depth. In addition to making products waterproof and water resistant, it might be necessary to advise users in the instructions for use, as well as by means of a warning label, to avoid exposure to water and moisture (see *Principle 79 - Add conspicuous warnings*).

IP code

The Ingress Protection (IP) code defined in ANSI/IEC 60529-2004, *Degrees of Protection Provided by Enclosures (IP code)*, characterizes the degree to which electrical equipment will resist intrusion by many elements, such as water, dust, and even body parts, such as fingers.

Example IP Rating

Resistance to solids
In this case, "3" means protection against a solid object greater than 2.5 mm.

IP 3 7

Resistance to liquids
In this case, "7" means protection from immersion in water with a depth of up to 1 meter (3.3 feet) for up to 30 minutes.[1]

Design guidelines for preventing fluid ingress

Exposure to fluids might not have an immediate negative effect. Rather, it might lead to corrosion over time that leads to electrical or mechanical failure. Below are some high-level guidelines for waterproofing products:

- Add a desiccant pack (which could pose a safety risk if consumed; see *Principle 93 - Label toxic substances*) to the product's package to absorb any moisture that might penetrate the main package.

- Set tight tolerances for device parts so that they fit together as tightly as possible.

- Make parts stiff to resist bending that could cause seams to separate and enable water ingress.

- When sealing parts using an O-ring, make the joining parts circular so the O-ring does not have to turn sharp corners.

- Do not squeeze gaskets too tightly, which can compromise their effectiveness at sealing.

- Allow for pressure equalization so that product components do not explode or crack in ways that enable water ingress.

Zippers with polyurethane coatings and special construction (e.g., rubber teeth) offer good water intrusion protection, which could make the difference between suffering from exposure and remaining warm and dry.

Use "TALLman" lettering

Principle

When displaying a word that looks or sounds like another word, use Tall Man lettering to make the word appear more distinct and less likely to be mistaken for its look-alike.

What is Tall Man lettering?

"Tall Man" lettering is a capitalization scheme that is often used in the presentation of medication names, but which can also be useful in other applications. The idea is to capitalize a contiguous string of letters in a drug name to make the overall name more visually distinct and less likely to be mistaken for similar looking and sounding names. Dobutamine and Dopamine are two such names given to drugs that have similar purposes (increasing heart output) but can have different effects on the body that could cause harm (e.g., severely increased heart rate) if administered erroneously due to a mix up. Tall Man lettering might also help differentiate similar words for individuals with vision impairments.

When using Tall Man lettering, capitalization should be applied to the letters that distinguish each name (e.g., "Dobut" and "Dop") to emphasize the differences. The distinguishing letters can be further emphasized by bolding them.

	Normal vision	Impaired vision
Normal lettering:	Dobutamine Dopamine	Dobutamine Dopamine
Tall Man lettering without bolding:	DOBUTamine DOPamine	DOBUTamine DOPamine
Tall Man lettering with the distinguishing letters bolded:	**DOBUT**amine **DOP**amine	**DOBUT**amine **DOP**amine

HydrALAzine Hydrochloride Tablets, USP

HydrOXYzine HCl Tablets, USP

Are my eyes playing tricks on me?

As mentioned on the previous page, Tall Man lettering involves capitalizing, rather than enlarging, distinguishing letters. However, the effect of capitalization, perhaps complemented by bolding, makes the capitalized letters look larger than the adjacent ones, which is part of the presumed benefit. Although the actual benefits of the practice have not been consistently demonstrated in controlled experiments—likely due to variations in methodology and study limitations[1]—the technique is now widely practiced and seems to be widely appreciated as part of a larger effort to make different drugs distinct.[2]

In fact, the use of Tall Man lettering is encouraged by the US Food & Drug Administration (FDA). In 2001, the FDA's Office of Generic Drugs (OGD) launched the "Name Differentiation Project" in an attempt "to minimize medication errors resulting from look-alike confusion."[3] As part of the project, the FDA identified many look-alike name pairs and then sent letters to the drug manufacturers encouraging them to revise their labeling to include Tall Man lettering. The table on the right shows several examples of established drug names and the name formatting recommended by the FDA.

Established name	Recommended formatting
Acetohexamide	acetoHEXAMIDE
Acetazolamide	acetaZOLAMIDE
Chlorpromazine	chlorproMAZINE
Chlorpropamide	chlorproPAMIDE
Daunorubicin	dAUNOrubicin
Doxorubicin	dOXOrubicin
Methylprednisolone	methylPREDNISolone
Methyltestosterone	methylTESTOSTERone
Tolazamide	TOLAZamide
Tolbutamide	TOLBUTamide
Vinblastine	vinBLAStine
Vincristine	vinCRIStine

Preventing errors, one Tall Man letter at a time

According to the Institute for Safe Medication Practices (ISMP), "One in every 1,000 medication orders in a hospital, and one in every 1,000 prescriptions in a pharmacy, have been associated with selecting the wrong drug while prescribing, transcribing, dispensing, or administering medication." And, "one of the key causes of these errors is drug name similarity."[4] In an ISMP survey involving 451 respondents:[2]

 87% felt that the use of Tall Man lettering helps reduce drug selection errors

 64% reported that Tall Man lettering has prevented them from dispensing or administering the wrong medication

So, although it is not mandated, Tall Man lettering certainly appears to have its benefits.

Childproof hazardous items

Principle

Products used by and around children should be childproofed to prevent dangerous events (e.g., choking, crushing, electrical shock) that could harm children.

The ABC's of childproofing

Childproofing involves creating a protective barrier between children and hazards. The protection might be physical, such as a cover over a hazard. Or, the protection might conceivably come from a young child's inability to perform tasks that older individuals can perform, such as those requiring reasoning, reading, fine motor control, and greater strength.

Childproofing the home

Sharp and pointed edges are an obvious hazard to children, which has given rise to a cottage industry of bumpers, gates, and other clever mechanisms serving to protect children from table corners, stairways, window cords, and many more seemingly benign features in the home. Risks to children posed by the home environment do not stop at bumping into sharp edges. Homes include things that can crush, burn, poison, electrocute, choke, and more. Risks are heightened by children's inability to recognize and avoid exposure to hazards (i.e., take precautions).

Common childproofing mechanisms include:[1]

Stair gates
Prevent children from falling down the stairs.

Bumpers
Cover sharp edges, such as table corners, to prevent children from injuring their heads.

Cabinet latches
Prevent children from opening cabinets that might contain cleaning supplies or other dangerous items.

Highchair straps
Prevent children from slipping down under the tray and being strangled. *(Model photo: Colourbox.com)*

Plug protectors
Prevent children from inserting their fingers or other objects into the wall outlet.

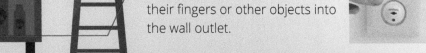

Other childproofing considerations

Arguably, childproofing should not be limited to children's products. Potentially hazardous products intended for use by adults at home (e.g., home-use medical devices and power tools) should have features that prevent use or tampering by children.

Unfortunately, childproofing is sometimes difficult to achieve without creating barriers to ease-of-use by the intended users. A childproof pill dispenser is one such example: while most children will not be able to open it because it requires strength, dexterity, and/or reading abilities that exceed them, some adults might also struggle to open it because it requires more manual strength and dexterity than a container with a simple twist-off top. Case in point, an older adult with arthritic hands might not be able to squeeze a specific part of the childproof cap with enough force to break the cap's grip on the container's neck.[2] This is why adults can obtain medications in containers that are not child resistant.[3]

In the US, childproof packaging has been required for almost a half century, having been mandated by the Poison Prevention Packaging Act of 1970.[4] In fact, 16 CFR 1700.20 - *Testing Procedure for Special Packaging* provides a detailed protocol for demonstrating that a pill bottle is child resistant.[5] According to the test's procedure, if a child can open the packaging or otherwise gain access to its contents over the course of at least one out of two 5-minute periods, it's considered a package failure. When it comes to unit packaging, the packages are evaluated by a 10-minute protocol: the unit packaging fails the test if a child can open or gain access to more than 8 individual units, or an alternative minimum number of units that could cause that child serious injury or illness.

One tragic type of accident involves highchairs. According to a report by the European health commission, "Estimates using EU Injury Database (IDB) data indicate that annually in the EU 28 Member States approximately 7,700 injuries to children 0-4 years of age involving high chairs are serious enough to require a visit to the emergency department. ...The most severe cases, those resulting in death, have occurred when children slipped down under the tray and were strangled. Most often, these children were either unrestrained or were restrained only by a waist belt (i.e. strap between the legs was not used)."[1]

Indicate expiration date

Principle

If a product is subject to aging and might be unsafe to use after a period of time, it should be labeled with a clear and prominent expiration date.

When it's over...

Many foods feature a "Sell-by...," "Best if used by...," "Enjoy by...," or "Expires on..." label. In principle, the inclusion of these dates helps consumers select fresh products and avoid those that might have already spoiled or will do so shortly. However, there is some concern that labeling inconsistencies lead to misinterpretations that, in turn, lead manufacturers, retailers, and consumers alike to dispose of perfectly good food.[1] That said, it's nice to know if a cup of yogurt expired a month ago.

Which products expire?

Expiration dates are not just for food—they can also can be found on manufactured products. A prominent example is an epinephrine auto-injector used to reverse allergic reactions, such as those that might occur among people who can have a life-threatening allergic reaction (called anaphylaxis) to a bee sting or eating peanuts, for example. Such devices can have a shelf life as short as 12 months because there is an increased chance that, after that point in the aging process, the formulation will deteriorate in ways that could affect the drug's effectiveness. Therefore, expiration dates are printed right on the devices. Consumers are expected to replace their auto-injectors before they expire. However, more recently customers have been advised to use expired devices in emergencies when nothing else is available because, in principle, receiving a reduced dose is still better than not receiving any medication.[2]

This might come as a surprise to some home owners, but smoke detectors also expire—typically after about a decade.[3] Although the expiration date might be printed right on the product, its inconspicuousness (the expiration date is typically located on the underside of ceiling-mounted smoke detectors) has likely caused many consumers to overlook it and, consequently, not replace their expired detectors on schedule. Fortunately, some of them start beeping when they expire (see *Principle 15 – Provide reminders*).

There are many more products that should carry an expiration date, but the matter is complicated by the use scenario. For example, mascara applicators may sit unopened on the store shelf for many months without concern. But, consumers are advised to throw them away as little as two months after first use because the moist, dark environment inside the tube is an ideal place for bacteria to proliferate.[4] According to a study published in *The International Journal of Cosmetic Science*, 79% of 40 mascara samples collected from the public were found to be contaminated with staph bacteria.[5] Some sort of "safe-use timer" might be beneficial, noting that many women will be reluctant to replace mascara applicators that still do their job, either due to the inconvenience or cost.

Save the date!

Expiration dates that really matter (i.e., those that are critical to safe product use) should be highly visible by virtue of their conspicuous placement and good legibility. Conversely, expiration dates that are essentially "hidden in the corner" and have poor legibility invite continued device use past the expiration date, often by someone who is unaware of the associated risk (or even that the product is expired).

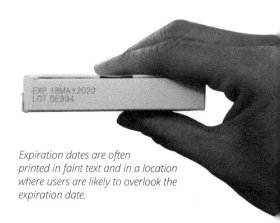

Expiration dates are often printed in faint text and in a location where users are likely to overlook the expiration date.

Tips for good expiration date design

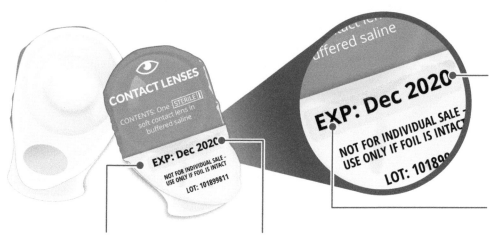

4. Spell out the month, or at least abbreviate the month to make the date distinct from other numerical information and avoid confusion. For example, "Dec 1, 2020" is clearer than "12-1-2020," which could be interpreted as December 1st or January 12th.

3. Label the expiration date (e.g., with "EXP") to reduce the likelihood that users misinterpret the date's purpose. For example, someone might misinterpret an unlabeled expiration date to be the product's manufacturing date.

1. Print the expiration date where it is likely to be seen during normal use. In this example, the expiration date is printed on the front of each individual contact lens packet and is highlighted yellow to stand out.

2. Use large (12-14 point or larger), bold text to make the expiration date stand out and to ensure the expiration date is legible when read at an arm's reach.

Make packages easy to open

Principle
Packages should not require special tools or instructions on how to open them.

Wrap rage

Have you ever struggled to open a new item that is packaged inside a molded plastic clamshell? Or, have you ever bought a new pair of scissors that is secured with wire ties to a cardboard backing and requires another pair of scissors to open the packaging?

"Wrap rage" is a term that describes the frustration caused by difficult-to-open packaging[1] (e.g., clamshells or plastic blister packs). Although some packages must be childproof (see *Principle 63 - Childproof hazardous items*), arguably, the rest should be easy to open. Saying that packages should be "easy to open" is different from saying that packages should "open easily." A door with a well-designed handle is easy to open, but the door should not open easily due to a person leaning against it. That is, unless it's the type of door that will open if people press against a horizontal bar—an innovation arising from deaths due to people not being able to escape fire from inside a room with a locked door and a surging crowd.[2]

Too many wrapping injuries

Difficult-to-open packaging can lead to more than "wrap rage." In 2004 alone, the US Consumer Product Safety Commission estimated that around 6,500 people went to the emergency room with plastic packaging-related injuries.[1] Some were caused by the sharp edges of half-opened packages. Others resulted from people trying to open the packaging with scissors, knives, can openers, or wire cutters and cutting themselves in the process.[3] In these cases, the packages almost certainly prioritized theft and tampering prevention over ease of opening and discounted the potential for the difficult-to-open packaging to cause injury. Difficult-to-open packages have created a market for products designed to make opening packages safer and easier.

The art of instructions

For convenience and sometimes for safety's sake, packages should not require special tools to open them. Most people are likely to agree that the opening process should not require undue strain or expose people to sharp edges. Moreover, packages should not require extreme tearing or pulling forces and then open abruptly, potentially propelling the contents toward the consumer's face, for example.

A simple and conspicuous instruction can frequently turn an epic struggle with a package into a satisfying or at least benign experience. Easy-to-locate instructions such as "TEAR HERE" and "CUT ALONG LINE" are often a big help. A package probably needs to be redesigned if the opening instructions have become a long essay.

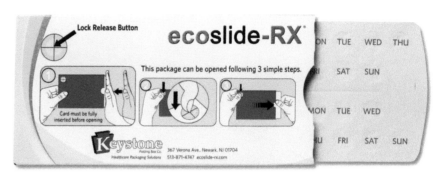

The ecoslide-RX package prevents pills from ending up in the wrong (i.e., children's) hands but ensures that users are still able to access the medication by providing its users with simple step-by-step opening instructions. (Photo courtesy of Packaging World)

Easy-to-open packaging solutions

Packages that have been designed to be opened safely without the need for special tools or opening instructions are perhaps even better. Here are some particularly usable opening solutions:

Tear strips
Tear strips can be opened with a quick, one-handed motion, enabling users to open the packaging easily.

Pull tabs
Pull tabs offer users a good grip, which makes opening the packaging easier. In some cases, pull tabs can also be used to reseal the package's contents.

Perforations
A line of holes in paperboard or a plastic film enables users to tear the packaging open without any extra effort or tools.

Cutouts
Cutouts provide an intuitive visual affordance, indicating where and how to open a package.

In 2008, Amazon started their popular "Frustration-Free Packaging" program to combat "wrap rage." Amazon focused on developing boxes that are easy to open (i.e., do not require cutting, slicing, or tearing), reduce opening injuries, are fully recyclable (e.g., are free from plastic clamshells), and are optimized for e-commerce (i.e., protect the product inside the packaging during shipment).[4]

Fight bad bacteria

Principle

Products used in healthcare environments, around food, or around vulnerable populations should fight against the spread of bad bacteria that can lead to bacterial infections and illnesses.

Bacteria: The good, the bad, and the really cool microscopic renderings...

Although scientific estimates regarding the ratio between microbes and human cells within the human body vary,[1] most of those within the scientific community have come to appreciate that the relationship between humans and microbes is largely symbiotic.[2] They need us. We need them. But, some microbes threaten a host's health. A few of the particularly nasty and unfortunately common bacteria—sometimes present in spoiled and contaminated food—are *Salmonella*, *Campylobacter*, *E. coli*, and *Listeria*.[3] People can be infected by these bacteria through ingestion, an open wound, and other routes.

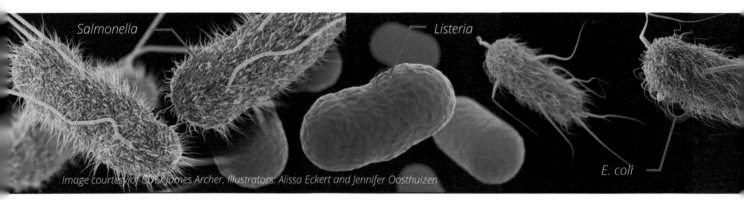

Salmonella

Listeria

E. coli

Image courtesy of CDC/James Archer, Illustrators: Alissa Eckert and Jennifer Oosthuizen

When a person contracts an infection, the body's immune system goes to work, possibly with the assistance of antibiotics. However, the medical community is increasingly concerned with the growing number and virulence of drug-resistant bacteria (e.g., Methicillin-resistant *Staphylococcus aureus*, also known as MRSA—a super bug). The Centers for Disease Control and Prevention (CDC) in the US is working diligently on a "National Strategy to Combat Antibiotic Resistance" based on four primary efforts:[4]

- Slow or prevent emergence of resistant bacteria and resistant infections
- Strengthen national surveillance efforts
- Develop rapid tests to identify and characterize resistant bacteria
- Improve collaboration with international organizations

*According to the CDC, antibiotic-resistant bacteria annually cause at least **2 million illnesses** and **23,000 deaths** in the United States.*[5]

How product designers can help reduce the chance of infection

Prevention of bacterial infection is an ongoing battle in numerous environments. Defensive tactics include:

Easily cleanable surfaces

Many products—especially those in hospitals and those designed for infants—can be designed with smooth, easily wipeable surfaces, devoid of crevasses that can harbor bacteria (see *Principle 2 – Make things easy to clean*).

Protocols

Disinfecting solutions, steam sterilization, and other techniques can be used, as appropriate, on regular schedules and according to recommended sterilization requirements (see *Principle 99 – Enable sterilization*).

Anti-bacterial dispensers

Antibacterial soap (or lotion) dispensers have proliferated. Public models often dispense lotion automatically (see *Principle 91 - Make it touch free*), rather than requiring users to press a bar touched by others with contaminated hands. Travel-size dispensing products have become very common as well, delivering bacteria-killing gels and sprays. One product even combines a dog poop bag holder and a sanitizer dispenser.

Reminders

We are all grateful when restaurants post reminders to employees to wash their hands after using the restroom. After all, keeping hands clean is one of the most important steps in avoiding the spread of germs.[6]

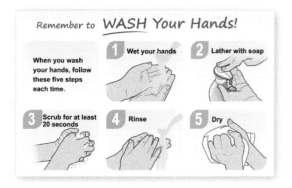

Remember to WASH Your Hands!

When you wash your hands, follow these five steps each time.

1 Wet your hands
2 Lather with soap
3 Scrub for at least 20 seconds
4 Rinse
5 Dry

Bacteria-hating materials

Consider selecting materials that can naturally resist bacteria. For example, you can use materials that incorporate EPA-registered alloys of copper that have been shown to resist bacterial contamination as well as kill bacteria that come into contact with the alloys.[7]

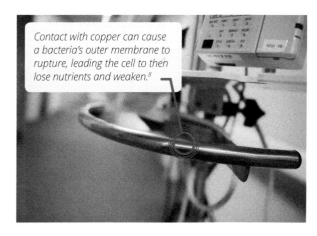

Contact with copper can cause a bacteria's outer membrane to rupture, leading the cell to then lose nutrients and weaken.[8]

Surface coatings

Another option is to coat objects/surfaces with a bactericide to prevent bacteria from surviving and/or spreading. One paint product often targeted toward hospitals earned the US Environmental Protection Agency certification for killing 99.9% of a wide range of dangerous bacteria including *staph*, *E.coli*, and *MRSA*. Reportedly, it kills the bacteria within about 2 hours of contact with the painted surface.[9]

The concept of bactericidal paint is similar to underarm deodorant that kills odor-causing bacteria such as *Staphylococcus hominis*.[10]

Number instructional text

Principle

Number instructional steps to help users perform actions in the correct sequence without overlooking critical steps.

Where should we start?

Instructions are often criticized for being poorly designed and poorly written. That is not to say that some instructions are not the opposite—composed to effectively guide people through the steps necessary to assemble, operate, or maintain the associated product.

By failing to follow instructions, people can commit use errors that might cause inconvenience or damage a product at a minimum, and at worst, cause significant injury. For example, failing to tighten a component on a power tool might cause it to fly off and strike the user or a bystander. Neglecting to put on protective gloves while handling a sharp blade could lead to a hand injury. Not following instructions to clean a device could lead to infection (see *Principle 2 - Make things easy to clean*). Not removing air from a syringe could lead to an air embolism.

Ikea is famous for removing text from their assembly instructions... except for numbered steps! If someone building a bookshelf performed the steps out of order, the result could be an unstable bookshelf that topples over and injures someone.

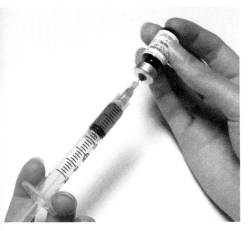

Numbered steps increase the likelihood that users follow the steps in the correct order and do not overlook critical steps, like removing air from a syringe before injecting.

Guidelines for composing effective instructions can and do fill textbooks. They address topics such as writing style (i.e., tone), topic order, page layout, graphical style (see *Principle 86 - Use graphical instructions*), and many more, including how to provide procedural guidance.

A key part of providing effective procedural guidance is numbering the steps. This is a way to: a) differentiate actions that must be performed from other kinds of information; b) make sure that readers follow steps in the correct order; c) make sure that readers don't overlook necessary actions; and d) help users stay on track while performing a lengthy procedure.

It's as easy as 1, 2, 3...

It's usually a good idea for the step number to stand out visually. This can be accomplished by making numerals large and/or a different color from surrounding content, integrating them with a graphic (e.g., header bar), and shifting them into the margin, among other ways.

When the total number of steps might seem daunting to users, it makes sense to group, or "chunk," some steps together and itemize them (e.g., Step 5 (a), Step 5 (b), Step 5 (c)). Arguably, and this is only speculation, mixing numerals with secondary letters (e.g., 5 (a)) as opposed to secondary numbers (e.g., 5.1) stands to make the substeps more memorable and less vulnerable to number transposition and confusion.

Another strategy is to chunk steps into subsections (e.g., A. Prepare device, B. Use device, C. Clean device) that have numbered steps within each subsection.

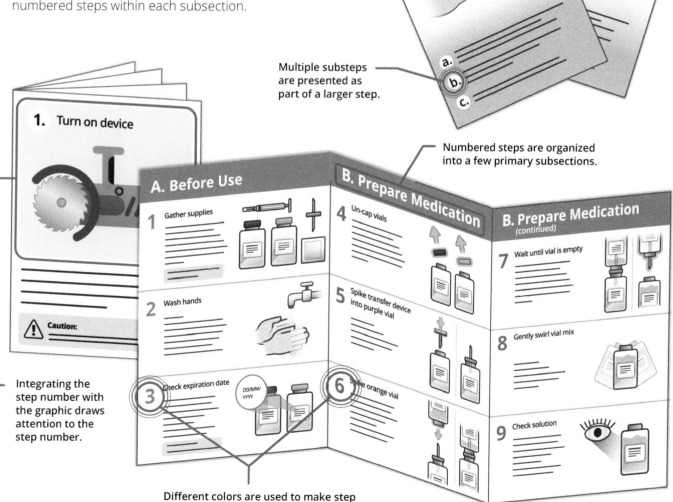

Large numerals make the step number stand out.

Multiple substeps are presented as part of a larger step.

Numbered steps are organized into a few primary subsections.

Integrating the step number with the graphic draws attention to the step number.

Different colors are used to make step numbers stand out and to differentiate subsections.

Prevent and expose tampering

Principle

Products that need to remain sealed until they are used should prevent unauthorized opening and/or warn users when someone has tampered with the product.

THE DAILY NEWS

57th year - No. 274	Friday, October 1, 1982	Final edition - $0.25

7 dead after tampering with Tylenol capsules

CHICAGO. Several people were killed by potassium cyanide poison that had been loaded into capsules in bottles of extra-strength Tylenol. Johnson & Johnson's McNeil Consumer Products (maker of Tylenol) recalled more than 31 million bottles.[1] This was a shocking and deadly case of product tampering that led companies to invent and implement various anti-tampering strategies and technologies and led to changes in the US tampering laws.

Tampering with consumer products is considered punishable and is a matter of criminal law. Title 18 of the United States Code § 1365 - *Tampering with consumer products* defines individuals who tamper with products as *"Whoever, with reckless disregard for the risk that another person will be placed in danger of death or bodily injury and under circumstances manifesting extreme indifference to such risk, tampers with any consumer product that affects interstate or foreign commerce, or the labeling of, or container for, any such product, or attempts to do so."[2]*

The consequences of tampering with consumer products include heavy fines and imprisonment. Unfortunately, no one was convicted for the murders of 1982.

tampering
ˈtam-p(ə-)riŋ

intransitive verb

1. a: to interfere so as to weaken or change for the worse—used with *with* • did not want to *tamper* with tradition
    b: to try foolish or dangerous experiments—used with *with*
    c: to render something harmful or dangerous by altering its structure or composition • was charged with *tampering* with consumer products

2. to carry on underhand or improper negotiations (as by bribery)[3]

How to prevent tampering

Solutions to discourage and reveal medication bottle tampering include adding a "safety seal" across a bottle's opening, adding a heat shrink seal around a bottle's top and cap, or adding a strip across a bottle's cap. Bubble packaging (i.e., blister packs) can serve the same purpose, exposing any tampering unless the criminal goes to great lengths to hide their tampering. The same applies to single dose packaging of pills in foil (tear) pouches.

Food packaging can protect against tampering using many of the same methods cited above. For example, yogurt containers typically have a peel-off foil seal.

   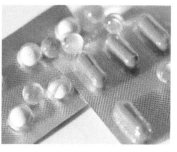

Many consumer products come with anti-tampering packaging to ensure that the products are not tampered with and are safe for consumption.

...or expose tampering

Specially designed packing tape will reveal if someone has opened a package. Undisturbed, the tape may have an opaque, unremarkable appearance as it does its primary job of sealing a cardboard box. However, once the tape is peeled back, it will leave a telltale message, such as "OPENED – VOID."

How to protect against other forms of tampering

Tampering of a different nature—changing computerized information in a manner that could be hazardous—can be combated by the use of passcodes, various forms of identity authentication (e.g., key card, electronic fingerprint, facial recognition), encryption, and access monitoring/tracking (see *Principle 19 - Make software secure*).

Sometimes, physical tampering can be deterred by using a simple lock and the threat of punishment to the perpetrator (see *Principle 26 - Incorporate a lockout mechanism*). For example, a warning and lock can be used to prevent unauthorized users from accessing an electrical unit.

 There have been some unusual and dangerous items found in products that have been tampered with, including mice, glass, liquid mercury, syringes, needles, cyanide, and many others.[4]

Put a cap on it

Principle
Put a cap or guard on sharp objects to protect against puncture wounds and cuts, maintain sterility, and prevent cross-contamination.

Cap it, cap it real good!

The world is full of sharp, pointy things. Besides natural ones such as cacti and porcupines, there are lots of man-made items that could cause puncture wounds were it not for a safety cap.

The need to cap syringes is obvious. Not only does a cap protect against puncture wounds, it also maintains sterility and prevents the needle tip from getting bent or blunted. The same can be said of the caps (or sheaths) on sharp knives, including those used in kitchens and hospital operating rooms (e.g., scalpels). In the case of disposable safety scalpels, the cap can be integrated into the handle and can slide forward to cover the blade, or the blade can retract into the handle[1] (like many box cutters), protecting against lacerations and infections. OK—maybe this is stretching the definition of a cap, but you get the point, noting that the boundaries between what constitutes a cap, sheath, and guard are not exact and clear-cut (pun intended).

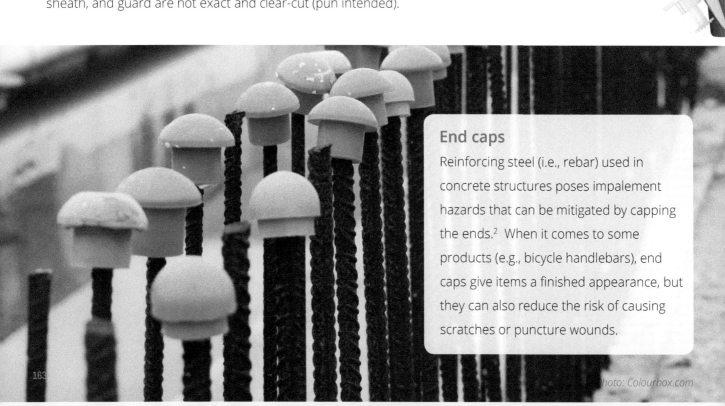

End caps
Reinforcing steel (i.e., rebar) used in concrete structures poses impalement hazards that can be mitigated by capping the ends.[2] When it comes to some products (e.g., bicycle handlebars), end caps give items a finished appearance, but they can also reduce the risk of causing scratches or puncture wounds.

Photo: Colourbox.com

Occupational hazards

Healthcare workers frequently confront operational hazards such as needlestick and sharps injuries. Each day, such workers expose themselves to deadly bloodborne pathogens via contaminated needles or other sharp objects.[3]

The World Health Organization (WHO) reported in the World Health Report 2002 that each year, 2 million out of 35 million healthcare workers experience percutaneous exposure to infectious diseases. In fact, needlestick injuries account for 37.6% of Hepatitis B, 39% of Hepatitis C, and 4.4% of HIV/AIDs in healthcare workers around the world.[4]

How should the cap be designed?

Clearly, a cap should be blunt rather than pointy. The cap should fit securely on the pointy item and not be vulnerable to unintended detachment. It helps for a cap to be integrated with or permanently connected to the item to help prevent users from misplacing the cap. Also, an integrated or attached cap makes it more convenient for the user to put it in place, such as with many syringes equipped with a needle guard, like the image on the right. Lastly, caps should be easy to remove. Although you don't want caps to fall off, a cap that requires a lot of force to remove can have a recoil effect whereby the user's hand snaps back towards the needle.[5]

Can you recap that for me?

The simple answer: no. The Occupational Safety and Health Administration (OSHA) prohibits recapping of contaminated needles unless recapping is required by a specific medical or dental procedure, or unless no alternative is feasible.[6] Although you might argue that recapping could prevent needlestick injury, the chances of someone accidentally reusing the contaminated needle, or stabbing themselves while trying to recap, poses a much higher risk.[7] If recapping must be performed, it must be accomplished by means of a recapping device which adequately protects the hands or by a properly performed one-hand scoop technique, as shown on the right.[8]

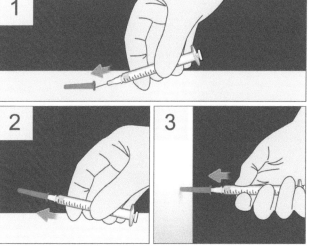

The one hand scoop technique: (1) Place the needle cap on a flat surface and insert needle into cap using one hand. (2) Scoop needle up into cap. (3) Using one hand, push capped needle against a firm, upright surface to "seat" cap onto needle.

Use hypoallergenic material

Principle

Products that contact the body should be made of hypoallergenic materials to reduce the potential for allergic reactions.

Allergic reactions

This book does not cover food products. So, concerns about allergic reactions to peanuts, eggs, and shellfish are out of scope. Bee stings are out as well. But, people can also experience allergic reactions to organic and synthetic materials that go into other types of products, and this is in scope!

"Allergic contact dermatitis" is the medical term used for a rash or irritation of the skin that occurs when allergens—substances that the immune system reacts to as foreign—touch your skin. It is estimated that atopic dermatitis—another type of skin-related allergy—affects up to 15-20% of children and 1-3% of adults worldwide.[1]

After exposure to an allergen, a severely reactive individual can go into anaphylactic shock and even die. Cause of death might be asphyxiation due to a closed airway or cardiac arrest due to a precipitous drop in blood pressure.

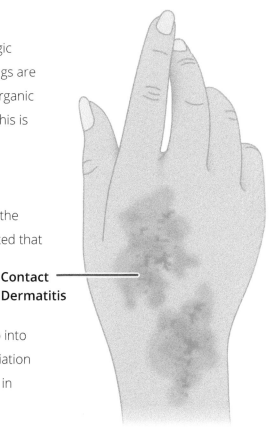

Contact Dermatitis

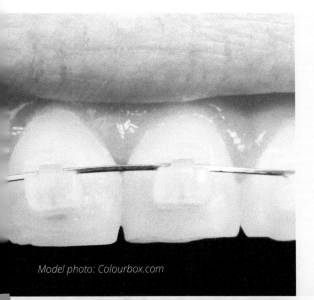

Model photo: Colourbox.com

Ceramic braces

People who have a nickel allergy will often have ceramic- or gold-plated braces instead of typical nickel braces. Nickel allergy is one type of metal hypersensitivity, an immune system disorder that affects 10-15% of the population and can produce symptoms such as swelling, rashes, or even pain due to contact with certain metals. According to some sources, up to 3% of men and 17% of women are allergic to nickel. Less commonly, 1-3% of people are allergic to cobalt and chromium.[2]

Allergic reaction-inducing materials and their substitutes

Below, we list a few materials that are likely to cause allergic reactions and identify substitutes that can reduce the potential for allergic reactions:

Latex

Some people react to natural latex rubber, a byproduct of the sap produced by the rubber tree. This has led to the development of many latex-free products, including protective gloves donned by food workers, medical workers, and even oil painters. Nitrile is a common replacement for latex. Notably, most latex paint does not actually contain natural latex. It contains synthetic latex, which does not trigger the same kind of allergic response that people have to natural latex rubber.[3]

Fiberglass

Many people develop a rash when exposed to fiberglass insulation, or at least find it mildly irritating. Manufacturers have responded by offering the material in an encapsulated form that reduces skin contact and, to some extent, the production of particulates that become airborne during handling and cutting.

Wool

Spoiler alert to wool lovers. Turn the page now. Wool has many remarkable qualities, including being a great insulator and fire retardant (see *Principle 9 - Make it fire resistant*). But, people can also develop a skin reaction due to contact with the natural material, which contains lanolin. Contact dermatitis can be mitigated by lining wool with another material where it might contact skin, such as by adding a fleece lining to a wool winter hat or covering a shirt collar with cotton broadcloth or silk fabric.

Wool hats are often lined with less-irritating materials like cotton or fleece.

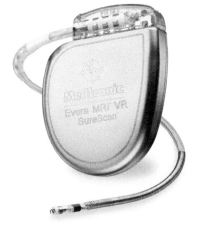

An implantable cardioverter defibrillator (ICD)—a device that tracks heart rate and delivers an electric shock when it detects an abnormal heart beat—is encased in biocompatible titanium.[4]

Metals

Some people have allergic reactions when their skin touches certain metals, particularly nickel (as described earlier), chrome, and copper alloys. Nickel is used in items such as zippers, watch straps, eyeglass frames, jewelry, pens, and coins. Chrome is used on watch cases, appliances, cookware, automobile components, and bicycle parts. Copper alloys are used in cookware, jewelry, and coins, as well as plumbing parts and countertops. Where contact dermatitis is likely, these metals can be replaced with more hypoallergenic materials such as niobium (used in medical implants), titanium, pure copper, stainless steel, silver, and gold. Another strategy is to use a coating that creates an invisible barrier, effectively sealing the metal item so it does not contact the skin.

Exemplar 7
Medication blister pack

Medication blister packs seal each pill in an individual compartment to help ensure that people take the right number of pills. Blister packs also have child-resistant outer packaging that prevents children from accessing and consuming the medication.

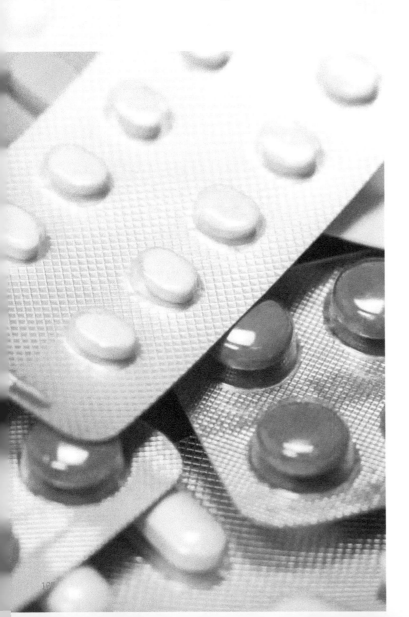

Use "TALLman" lettering
Principle 62 - pg. 149

Tall Man lettering helps prevent users from mistaking the medication name with another one that might look or sound similar.

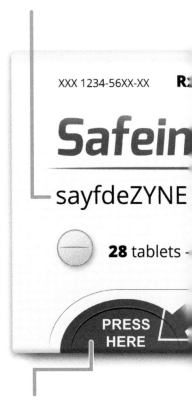

XXX 1234-56XX-XX   R

Safein

sayfdeZYNE

28 tablets

PRESS
HERE

Childproof hazardous items
Principle 63 - pg. 151

Outer packaging "lock" mechanisms add enough complexity to the opening process to prevent children from accessing the medication.

Use hypoallergenic material

Principle 70 - pg. 165

Hypoallergenic materials used in the medication's packaging reduce the likelihood that its users will suffer an unpleasant allergic reaction.

Make packages easy to open

Principle 65 - pg. 155

Cutouts in the outer packaging inform users where and how to comfortably grip the medication pack to remove it from the outer packaging.

Indicate expiration date

Principle 64 - pg. 153

Expiration dates—often stamped on packaging following the abbreviation "EXP"—inform users of the date beyond which users should not take the medication due to the potential for known and perhaps unknown side effects.

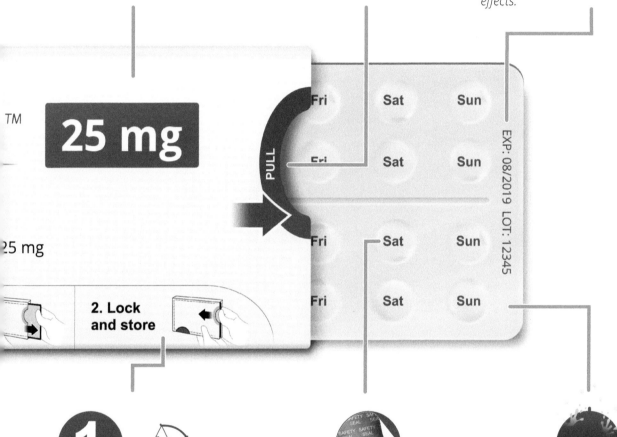

Number instructional text / Use graphical instructions

Principle 67 - pg. 159 / Principle 86 - pg. 201

Numbered, graphical instructions help users quickly identify the proper sequence of steps required to open the packaging, avoiding a potentially harmful delay in taking an emergency medication.

Prevent and expose tampering

Principle 68 - pg. 161

A broken foil seal indicates that someone (or something) might have tampered with or disturbed the medication in some manner.

Prevent fluid ingress

Principle 61 - pg. 147

Each capsule's individually sealed compartment protects it from moisture and exposure to other contaminants.

Manage and stow cords, cables, and tubes

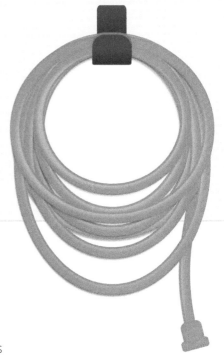

Principle
Products should enable users to store and organize cords, cables, and tubes to prevent tripping hazards and ensure they do not get lost or damaged.

Untangling cables
Some products can develop a Medusa-like appearance, festooned with cords, cables, and tubes in cases where there is no place to put them. This represents a design failure in view of the many ways that device accessories can be stowed out-of-the-way.

In lieu of proper storage options, cords, cables, and tubes can fall to the ground and become tripping hazards. Furthermore, these accessories might become tangled, frayed, and even torn away if they hang loose or are carelessly wrapped around the associated devices or any sharp edges.

If cords, cables, or tubes become unavailable or inoperable, a device might not be able to perform a safety-critical function. For example, if a patient monitor/defibrillator's electrode or pad cables are damaged, a paramedic might be unable to monitor a patient's heart rhythm or deliver an electrical cardioversion (i.e., a shock to restore regular rhythm).

Organizing cables
Storing cords, cables, and tubes can be as simple as equipping a product with storage baskets or compartments. Power cords can be organized using recoiling mechanisms or brackets. Smaller diameter cables and tubes may be coiled and placed in covered bins or bags, or they can be gathered in a single channel that keeps them in an optimal, safe position.

Recoiling mechanism

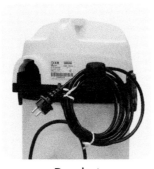
Bracket

Basket / compartment
Photo: Colourbox.com

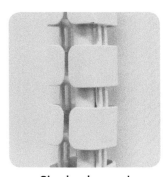
Single channel

Recoiling mechanisms

Some products, such as vacuum cleaners, feature a recoiling mechanism that automatically retracts and coils the power cable inside the product when the user presses a button. In addition to reducing strain on the user by automatically coiling the cable, it ensures the cable remains organized and undamaged inside the product. Furthermore, the coiling mechanism reduces the likelihood that the cable becomes a tripping hazard when the product is not in use.

As a user pulls out the power cable, a spring inside the recoiling mechanism is compressed. When the user presses the release button, the spring's energy is released, pulling the power cable back inside the product.

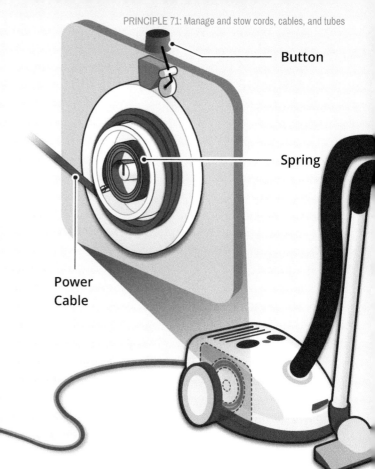

Button

Spring

Power Cable

When to use brackets versus recoiling mechanisms

The relatively high cost and weight of a recoiling mechanism might make brackets a preferred solution in certain cases. But, people are well-served by recoiling mechanisms in cases where they are practical because of their superior ease and speed of use. In turn, efficient and proper power cord storage enhances user safety.

Beyond household devices, retractable power cords can also be found in medical equipment, such as endoscopy carts, telemedicine carts, and stretcher beds.[1] We also find instances of bracket-like mechanisms in healthcare settings. Consider a situation in which it is critical that a tube stays in place, such as for an intubated patient on a ventilator. In these cases, a tube-positioning device might be needed (see image on the right).[2]

When designing brackets that will store cords and cables, ensure they facilitate complete and effective accessory winding. Too often, brackets are undersized, which complicates secure winding. Additionally, some brackets are crude elements with sharp edges that can pose a cutting and abrasion hazard to users' hands.

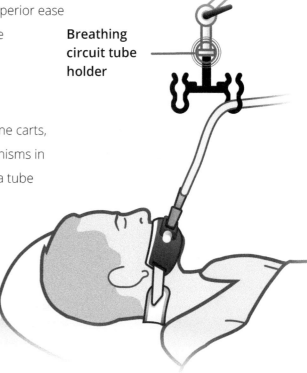

Breathing circuit tube holder

A tube positioning device keeps an intubated patient's breathing circuit secure and relieves stress/strain at the patient connection point.

Indicate radiation exposure

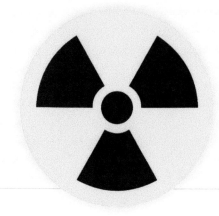

Principle
Use warning labels, alarms, dosimeters, and personal protective equipment to protect users from radiation exposure.

Warn against radiation exposure

When it comes to avoiding harm due to radiation, forewarning is a good first step because radiation poisoning can occur quickly and to a devastating extent before someone is aware or shows symptoms.

Places where routinely higher levels of exposure are possible include nuclear submarines and nuclear power plants, dental clinics, cancer treatment centers, and operating rooms.

There are also natural environments in which the minerals emit a considerable amount of radiation. A beautiful beach in Guarapari, Brazil turns out to be a scenic hotspot where a resident's annual radiation exposure reaches more than 8 times the annual limit set for nuclear power plant workers: 175 versus 20 millisieverts (mSv).[1]

People can wear radiation suits for protection from some types of radiation where it helps to prevent skin contact with radioactive particles, inhalation of such particles, and the penetration of certain types of low-level radiation.

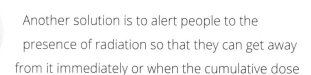

1000 microsieverts = 1 millisievert. Geiger counters make a ticking sound whenever significant radiation is observed.[2]

Another solution is to alert people to the presence of radiation so that they can get away from it immediately or when the cumulative dose approaches a recognized safety limit. Example solutions are signs and (product) labels that make a person aware of the risk before entering a potentially radioactive area or before interacting with a radiation-emitting product.

While radiation can certainly be harmful, people sometimes overestimate the risk.[3] Consider the Fukushima nuclear meltdown resulting from the March 11, 2011 tsunami. The meltdown led to panic in Tokyo (Japan's capital), roughly 240 km away from the nuclear reactor. As a result of the meltdown, radiation levels of 400 mSv per hour were emitted near the nuclear plant.[4] The media reported radiation levels more than 20 times higher than normal in Tokyo.[5] For a few hours, radiation levels in Tokyo went from 0.126 mSv to 2.5 mSv—about half the dose you would normally receive from an abdominal x-ray.[6] Still, it makes sense to minimize unnecessary exposure.

Wearable radiation indicators

Some people might have to spend a considerable amount of time in environments where radiation exposure is routine, intermittent, or possible (in an accident scenario, for example). Tracking how much radiation they have been exposed to is key to preventing radiation sickness. A common wearable solution is a dosimeter (often called a radiation badge). These badges have a detection area with a radiation-sensitive film that changes color when exposed to radiation.[7]

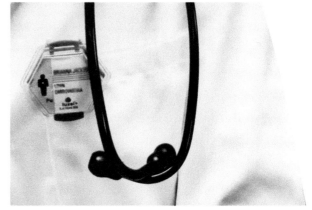

Healthcare professionals exposed to X-ray or Gamma emitters (e.g., radiologists) can wear dosimeter badges to monitor their prolonged exposure to radiation.

Some healthcare workers also wear a ring dosimeter if they expect that their hands might be most likely to be exposed to radiation. Thermoluminescent dosimeter (TLD) rings visibly light up when exposed to radiation.[8] Some dosimeters have been marketed to pregnant women to alert them to the presence of radiation that could have a significant effect on a developing fetus. One of the advantages of dosimeter badges is that they can be clipped to a shirt collar, carried in a pocket, or tied to a bag or case for ease of portability.

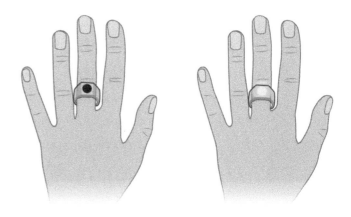

Dosimeter rings measure radiation exposure to a user's hands. When ionizing radiation enters the ring, its material (lithium fluoride) heats up and becomes luminescent, warning the user that radiation exposure is approaching an unsafe limit.

Dosimeters are not limited to badges and rings. They also take the form of watches, a clip-on devices with digital displays, and gadgets that connect to mobile phones. These more advanced dosimeters provide accurate measurements of specific environments that can be immediately stored and processed together with geographic coordinates to quickly map radiation levels in a particular zone.[9] Although one can resort to using a classic Geiger counter (image on the right), it is not always a handy tool due to its bulky size and need to carry it around.

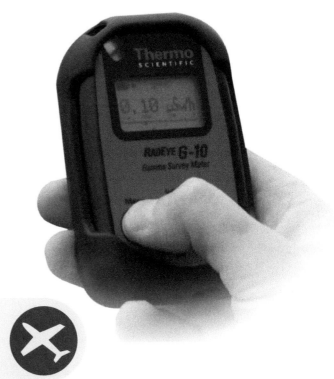

Flying from New York to Los Angeles exposes passengers to an extra 30-40 microsieverts. This is 2,500 times the amount of radiation passengers are exposed to when scanned by a full-body X-ray scanner at the airport.[10]

Shut off automatically

Principle
Products that could pose a safety risk if left running or unattended should automatically shut off after an appropriate duration.

Shutting down in 3, 2, 1...

Most would agree that you cannot rely on people to exercise precautions or remember to take proper actions. That is why certain devices and equipment shut off automatically after a certain amount of time. You see such features built into many devices that could immediately cause harm due to foreseeable misuse and accidents (e.g., coffee makers, clothes irons).

In some cases, it is safest for devices to automatically reverse a prior user action (e.g., suspending an alarm, disengaging a speed limiter, placing a mechanism into a manual mode), which may be considered another type of shutoff. When doing so, it is important to indicate the mode change to the user in a conspicuous manner to alert the user of the new mode and its new functionality.

Some automatic shutoffs are under software control wherein the software intervenes and shuts off the product after a period of inactivity. Some detect motion, such as an item entering an unsafe zone. So-called electric eyes disable equipment when something is detected in a scanned field or interrupts an energy beam.

Example:
Hospital sterilization system
The Getinge Air Glide System (AGS) automatically sterilizes items without needing supervision. The system moves carts in and out of sterilizers within a laser-based perimeter; if the laser beam is broken, the system stops moving, preventing potential collision with users.[1]

When should I incorporate an automatic shutoff feature?

Consider including an automatic shutoff feature in products that contain or emit:

Heat

Many products produce heat. Incorporate an automatic shutoff if the product topples over or reaches a certain high temperature for a period of time.

Sample risk: Unattended portable heater could be knocked over by a restless dog and start a fire.

Radiation

Some products emit radiation during operation. Incorporate an automatic shutoff to prevent user exposure.

Sample risk: Microwave oven could radiate a cook if it does not immediately shut off when the cook opens the oven door.

Gas or liquid

Products that carry toxic or flammable gas or liquid should feature an automatic shutoff when the fluid reaches a certain level or when there is potential leakage.

Sample risk: Gas tank could overflow and flood a gas station with highly flammable fluid.

Moving parts

We interact with many machines and systems that have moving parts. Incorporate an automatic shutoff to halt moving parts when a user is in the path of contact.

Sample risk: Automated revolving door could pin and crush someone who falls while entering the door.

When are automatic shutoffs inappropriate?

An automatic shutoff feature might not be suitable for certain products. For example, you likely wouldn't want a ventilator providing a patient with oxygen to automatically shut off for any reason. Make sure that the automatic shutoff does not prevent the delivery of life-saving treatment.

Tips for preventing inadvertent automatic shutoffs:

✔ Display a countdown to the shutoff (e.g., "System will shut down in 30 seconds").

✔ Provide both visual and audible cues to alert users of the looming shutoff.

✔ Enable users to cancel the automatic shutoff (perhaps for a set time, like a snooze feature).

Example: automatic braking systems

According to the Insurance Institute for Highway Safety (IIHS), vehicles with automatic braking systems have lower rates of rear-end crashes than those without such automatic systems.

Specifically, according to IIHS research conducted between 2010 and 2014, equipping cars with both auto-braking and warning systems led to a 39% decrease in rear-end crashes and a 42% decrease in rear-end crashes with injuries, which translates to a 12% decrease in crashes and a 15% decrease in injury crashes overall.[2]

Evacuate smoke

Principle
Products that produce dangerous smoke particles or fumes should incorporate a means to evacuate them.

Why is smoke dangerous?

Smoke is a byproduct of burning something, often in an incomplete manner due to the lack of a sufficient oxygen supply. It is typically a mixture of small solids, gases, and vapors (e.g., mists). Gases, smoke, or vapors that smell strongly or are dangerous to inhale are called fumes.

Smoke can be harmful because it is hot, contains dangerous chemicals and biohazards (e.g., bacteria, viruses),[1] and deprives someone of the normal amount of oxygen. Accordingly, inhaling the smoke produced by products that cut, cook, cauterize, melt, and otherwise heat things is not good for you. That is why some products incorporate a means to get rid of the bad byproducts.

Electrosurgical fumes

The Occupational Safety and Health Administration (OSHA) estimates that each year over 500,000 healthcare workers are exposed to smoke generated by electrosurgical devices that use lasers and electricity to burn tissue.[2] Traditional surgical masks do not offer enough protection against the smoke, which carries a risk of biocontamination (e.g., transmission of HIV, HPV, Hepatitis B) and poses a risk of cytotoxic, genotoxic, and mutagenic effects.[3]

Inhaling surgical smoke (i.e., smoke plume) can cause mild to potentially dangerous side effects including dizziness, headache, eye and respiratory irritation, nausea, vomiting, skin irritation, and potential carcinogens. Not good!

Worldwide, there are various safety guidelines in place for protecting users from smoke inhalation during electrosurgical procedures:[2]

- **Australia:** ACORN Standard S20
- **Canada:** CSA Z301-13
- **Denmark:** AT-instructions 4/2007 and 11/2008
- **Germany:** TRGS 525, 8.1
- **UK:** MHRA DB2008(3) and AfPP 2007 Standard 2.6
- **US:** OSHA General Duty Clause

Smoke evacuators

Smoke evacuation invariably involves creating a partial vacuum near the source of smoke, thereby drawing the contaminated air away from people in its vicinity. Sometimes, the smoke passes through a duct to some other location where it is safer to vent it to the atmosphere. In other cases, the smoke is passed through a fume evacuator that makes use of replaceable filters that remove dangerous particles and odors and blows the clean air back into the room.

A hood uses an exhaust fan to draw away smoke before cleaning the air and exhausting the smoke outdoors.

Smoke evacuators are a common accessory to surgical devices that cauterize tissue. Getting rid of the smoke not only likely makes the work environment clearer and less smelly but also protects members of a surgical team from biological risks of inhaling it. Smoke evacuators are also used in metalworking jobs, such as those involving welding and soldering. Typically, the workspace will contain a standalone or accessory device that acts like a vacuum cleaner, drawing away the smoke plume.

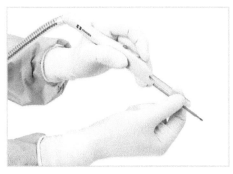

On-tool extraction, like the small smoke evacuation tubing integrated into the surgical tool, is effective because it extracts smoke closest to the source while giving the surgeon freedom of movement. (Photo courtesy of Medtronic Inc.)

Design considerations

When designing a smoke evacuator, there are several factors to consider:

- **Rate of suction**
  The rate of suction should be faster than the rate of contaminated air being emitted from the smoke-producing device.

- **Shape, size, location, and position**
  The shape, size, location, and position of the smoke evacuator all matter. The evacuator should be as close to the smoke source as possible but should not interfere with the work at hand.

- **Material**
  The evacuator should be made of a material that can withstand exposure to the dangerous chemicals and biohazards so that users are not at risk of the evacuator breaking down.

- **Running indicator**
  A smoke evacuator should clearly indicate whether smoke evacuation is active to ensure that a non-functional evacuator doesn't go unnoticed.

- **Expected noise level**
  A smoke evacuator should not produce so much noise that it violates environmental standards or tempts users to turn it off due to the annoyance it causes.

- **Type of filters**
  A smoke evacuator should contain a filter that is appropriate for the application. There are filters that trap particles forced through a fine mesh (e.g., HEPA filters), filters that use an electrostatic charge (i.e., ionization) to attract and pull particles out of the air stream, and filters that use activated carbon to chemically bond with the particles, thereby trapping them (see *Principle 27 - Eliminate or limit toxic fumes*).

- **Filter change indicator**
  Filters have a limited lifespan. Therefore, evacuators should indicate when filters need replacing for effective smoke removal and filtration to continue.

Protect against electric shock

Principle
Incorporate electric shock prevention mechanisms that protect against painful or potentially life-threatening electric shocks.

Shocks to the system
Alternating current (AC)-powered devices can become dangerous if a live wire (e.g., one with frayed insulation) contacts a conductive surface (e.g., an appliance's metallic case). One way to protect users from getting shocked by a faulty device is to build in a ground-fault circuit interrupter (GFCI) mechanism.[1]

GFCIs rapidly cut power to a device when the difference between the amount of current flowing to and returning from a piece of equipment exceeds a given threshold, indicating a ground fault. Someone holding onto a device subject to a ground fault could be electrocuted, burned, or (fatally) injured.[2]

Grounding electric circuits
Electrically grounded circuits also provide protection because they add a third wire—the ground wire—that provides a low-resistance path for electricity to flow safely to the ground in the event of a short circuit.[3] When current flows through a ground wire, it trips a breaker, thereby protecting the user. However, some AC receptacles, particularly older models with two prongs, rather than three, are not grounded.[4]

Many blow-dryers and curling irons incorporate GFCIs because the appliances are typically used near water (e.g., a filled sink). Water provides a low-resistance path and the opportunity for electricity to flow from a faulty device to the sink's metal plumbing, with the user caught in the middle.

Tips for avoiding electric shocks:

 Don't use electric appliances near water or while touching faucets or water pipes.

 Don't attempt to fix electrical appliances on your own, even if they appear simple (see *Principle 45 - Require professional maintenance and repair*).

 Don't use appliances that have damaged plugs or cracked wires.

 Unplug appliances when they're not in use.

Isolate electric wires or appliances

Another way to protect against electric shock is to double-insulate a device so that a live wire cannot transfer electricity to the outer casing that users touch. For example, an appliance manufacturer can put a second layer of insulation around already-insulated wires. Or, a device's casing can be made from a nonconductive material—typically plastic—that does not transmit electricity, thereby creating a second insulating layer.[5]

In Europe, double-insulated products that do not have a grounding (i.e., earth) wire must include the IEC (International Electrotechnical Commission) double-square symbol on its label. The two squares represent the two layers of insulation that protect the user from electrical shock.[6]

An electric hand saw's label indicates it is double-insulated.

Shocking insights

- Wiring can fray due to contact with sharp edges that can wear down or nick the protective insulation. Also, wire insulation can melt due to proximity to heat sources or due to the wire itself overheating.

- One way to protect an electrical cord from damage is to build in strain relief by providing flexibility or additional play where the cord exits the device. Another is to provide a way to coil or retract the wire to get it out of harm's way (see *Principle 71 - Manage and stow cords, cables, and tubes*).

- It takes only 5 mA (0.005 A) of current leakage from the hot wire to the ground to trigger a GFCI.[7]

- Death due to electric shock is not the result of high voltage, but rather of high current.[8]

- Shocks above 10 mA are capable of causing muscle paralysis.

- Electric shock can kill. Risk of death is highest when the current is between 100 and 200 mA because current in that range causes the heart's ventricles to twitch in an uncoordinated manner, leading to cardiac arrest. Counterintuitively, shocks above 200 mA are less likely to kill because the muscle contractions are so severe that they can force the cardiac muscles to clamp, thereby preventing the heart from going into ventricular fibrillation.

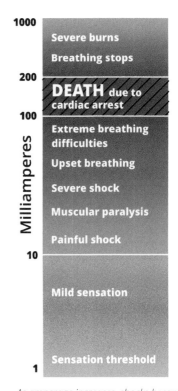

As amperage increases, shocks become more severe, with the potential to kill in the 100 - 200 mA range. Surprisingly, chances of survival increase when current is above 200 mA.[9]

Flash at an appropriate rate

Principle

Use an appropriate flash rate (i.e., duty cycle) to capture a user's attention in an urgent situation.

Why flash?

Things that flash are likely to draw attention better than equivalent things that are steadily displayed. This explains why police cars are equipped with flashing lights. Human beings are sensitive to changing stimuli.

However, not all flashing indicators are created equal. Some flash at a rate that seems "about right," while others flash too quickly or slowly. A rate that usually seems "about right" is 1-2 Hz, meaning 1-2 times per second.[1] This rate helps to ensure that people glancing briefly at the light notice that it is flashing.

Flashing speeds

One problem to avoid is flashing an indicator too slowly (e.g., an on/off duty cycle that repeats every 3 seconds). For instance, someone who is briefly scanning a control panel could potentially miss a critical indication because the light is off for the full second that he is looking at the indicator.

On the other hand, flashing too rapidly causes a different problem. Indicators that flash too rapidly can be annoying, as if the light is flickering due to a component failure rather than cycling rapidly to communicate a special meaning. If a light flashes even faster—around 60 Hz or higher—people will perceive it to be constantly illuminated, a phenomenon called persistence of vision.[2] This is how we can view movie frames projected one frame at a time and not notice any flicker.

Early films were projected at a rate of 16 frames per second (fps), which caused a noticeable "flicker" between light and dark as the images came and went. "Flicker," which evolved to "flick," became the slang term for a movie.[2]

Flashing duty cycles

Rather than engineer or program an indicator to be on and off for equal lengths of time (i.e., a 50% duty cycle), it can be advantageous to set the light to stay on longer than it is off. An indicator flashing at 2 Hz (i.e., full cycle every 500 ms) with a duty cycle of 3:1 would be on for 375 ms and off for 125 ms. This rate and cycle might cause an indicator that illuminates and extinguishes only briefly to appear to pulse rather than flash.

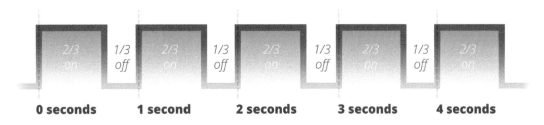

3:1 duty cycle graph displaying when the indicator will flash on and off.

Modern lighthouses appear to pulse as a rotating beacon light rotates within a lens that concentrates the projected light. A repetition rate of once every 5, 10, or 15 seconds (i.e., 12, 6, or 4 times per minute, respectively) is common. Some lighthouses generate burst patterns by emitting 2, 3, or 4 flashes in quick succession and then repeating the burst pattern every 10 to 20 seconds.[3]

Flash it like you mean it

Sometimes, flashing indicates that something is in progress or acts as a call-to-action (e.g., press the blinking power button to activate the machine). Other times, it indicates that something is wrong. The indicator's color can make the difference. A red flashing light seems to be used more often to indicate something bad is happening. However, color-coding within and between specific industries can be inconsistent. That is why a flashing red light can mean both that a dangerous fire has been detected, but also that a battery pack is properly recharging and is not yet fully charged.

A fire alarm in a school building that flashes red when actively alarming.

Flash safely

Exposure to flashing lights with certain visual patterns or at specific frequencies can trigger seizures in about 3% of individuals who have epilepsy. The condition is called "photosensitive epilepsy." To protect against seizures in such individuals, the Epilepsy Foundation recommends keeping flashing rates below 2 Hz and including occasional pattern breaks. According to one source, lights flashing at frequencies between 5 and 30 Hz are most likely to trigger such seizures.[4] Patented in the early 1900s, turn indicators in the US are required to blink at a rate of 60-120 blinks per minute (1-2 Hz)[5]—a rate that is unlikely to trigger a seizure.

Prevent glare and reflections

Principle
Design products and systems to prevent glare and reflections, thereby ensuring that users can see things clearly and accurately.

What's the issue with glare?

Glare (i.e., reflected light) is rarely a good thing—it interferes with seeing things, which, in turn, can lead to accidents. For example, glare can cause people to run into unseen objects, which is exactly what happens in many automobile and downhill skiing accidents.[1,2] Glare obstructing on-screen information can also be hazardous. As an example, glare on a computer screen could cause a clinician to misread a patient's lab result, resulting in an incorrect diagnosis.

Glare from the sun or reflections from a shiny dashboard can prevent drivers from seeing the road and surrounding cars clearly (or at all).[3]

Sunlight reflecting off the water can make it very difficult to see. That's why sailors often wear sunglasses with polarized lenses.[4]

Glare on critical cockpit displays can make it difficult for pilots to acquire information quickly and accurately.[5]

Glare can take the form of an overwhelming amount of light flooding one's field of vision, as when driving in the direction of a rising or setting sun. Glare can also take the form of light reflecting off of glassy surfaces, such as computer monitors, mirrors, and windows.

Hint: You can make a display that is vulnerable to glare a bit more legible by turning up the brightness (i.e., luminance) so that displayed content "powers through" the glare to some degree. In other words, you can improve the luminance ratio between the information and the interfering light.[6] You can also make critical information large and increase its contrast to maximize legibility in the face of glare (see *Principle 48 – Make text legible*).

Tips for reducing or preventing glare

To prevent glare and reflections:

- Avoid the use of shiny materials and finishes that can cause glare. This is a tip that some car manufacturers could take to heart, eliminating the bright instrument bezels and glossy surfaces that are common to many dashboards. Drivers don't want to be "blinded by the light," as goes the song performed by Manfred Mann and written by Bruce Springsteen.

- Ensure displays are adjustable so that users can position them to prevent or reduce glare caused by overhead lights or streaming sunlight.

- Add anti-glare and anti-reflection coatings or films. This is especially helpful for devices commonly used in bright lighting conditions like direct sunlight.

- Give surfaces—such as a glucose meter's screen—a matte finish that causes greater light dispersion and reflects less light in a particular direction. But, keep in mind that a matte finish can reduce image quality to a certain degree.[7]

We care about glare! That's why we printed this book on non-reflective matte paper, so that you can enjoy reading glare-free text on a sunny beach or in your well-lit office. You're welcome.

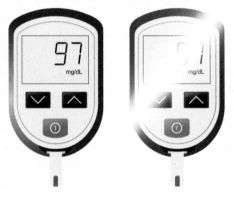

A blood glucose meter without glare (left) reads 97 mg/dL. However, the same meter with glare (right) appears to read 57 mg/dL.

A case study in glare

Trying to read a blood glucose meter when there is glare on the screen is not only frustrating, but it can also be hazardous. For example, a blood glucose meter might display 97 mg/dL—typically within a "normal" range. However, glare could cause the value to appear to be 57 mg/dL—a low reading, often indicating hypoglycemia. Thinking his/her blood glucose is low, a user might consume carbohydrates, which could raise their already-normal blood glucose to a dangerously high level.

I can see clearly now the glare is gone

Most people are familiar with polarized lenses used in sunglasses. Polarizers are a type of filter that blocks reflected light beams, while still enabling light beams coming straight from their sources to pass through to the viewer's eyes.[4]

Anti-reflective coatings on eyeglasses reduce the amount of light that bounces off the lenses and toward the wearer's eyes, as well as toward others.[8] That is why photos of people wearing glasses with anti-reflective coatings tend to look better, or at least you can see their eyes through their glasses.

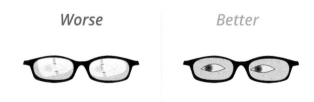

Worse *Better*

Guard against sudden static discharge

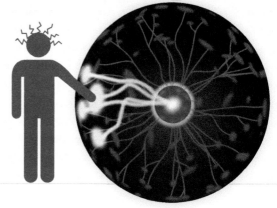

Principle

Products should prevent or minimize static buildup and discharge that could cause a fire or explosion.

What is static discharge?

Static electricity makes socks cling to pants exiting the dryer. A discharge (low-amperage spark) can occur when someone walks across a carpet and then touches a light switch, for example. Patting a cat's fur on a dry day can produce a light show of tiny sparks...and probably get the feline's attention.

But, static electricity can also be dangerous. A spark resulting from a buildup of static electricity can ignite gasoline vapors emanating from a gas tank fill port, potentially leading to a conflagration (or, in other words, a large fire!). The spark can come from the hand of a customer who (1) started fueling by putting the nozzle in the hold-open position, (2) subsequently got back into the car (perhaps to stay warm) and slid across the seat in a manner that generated a static charge, and (3) finally got out of the car to finish pumping. To address this risk, as well as reduce air pollution, many fuel nozzles[1] and newer cars[2] are engineered to collect what would otherwise be escaping gasoline vapors. Also, many pumps have warning signs intended to inform users of the static electricity risk. People can protect themselves further by either not re-entering the car while refueling or grounding themselves when emerging from the car, perhaps by touching the car's metal door before touching the nozzle. Some pumps lack hold-open clips by design, requiring users to continuously squeeze the nozzle handle while refueling and discouraging them from re-entering their cars until they've finished.[3]

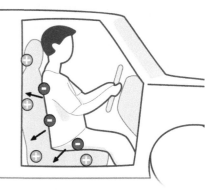

By rubbing against the car seat, the driver exchanges charges with the seat and becomes negatively charged.

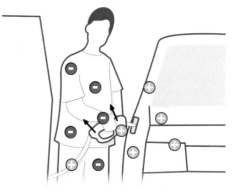

When the driver touches the nozzle, the charges try to balance out and make the driver neutrally charged again.

The suggestion that a ringing mobile phone can spark a fire during refueling at a retail gas station has been dismissed by some as an urban myth.[4]

This exchange of charges is called static discharge and can cause sparks that can ignite gas fumes.

How to prevent static discharge

There are multiple ways to prevent unsafe static buildup and discharge:

Grounding

Helicopters can develop a considerable static charge due to the rotor blades sweeping through the air. To deal with the hazard, the "Static Man" on a support team has the job of grounding the helicopter using a grounding rod.[5]

Similarly, tank trucks that contain large quantities of chemicals can generate static electricity during loading and unloading. This could lead to electrostatic sparks with enough energy to ignite a vast range of combustible gases, vapors, and dusts.[6] This hazard has given rise to special grounding devices that draw static charge away from the charged object and monitor for an unlikely, unsafe buildup.

The "Static Man" holds a grounding rod to discharge the static electricity generated by the rotor blades.

Many homes feature lightning rods that direct energy from a lightning strike straight to the ground.

Electricity distractors

The ultimate static discharge is a lightning bolt. People can protect their homes with lighting rods that redirect the energy from a lightning strike to the ground. Homeowners can also protect equipment that might catch fire by installing surge protection devices. And, people can protect themselves by knowing when to get indoors with the help of advanced warning from lightning detection devices that sense the electromagnetic emissions from lightning bolts.

Flame arrestors

Under the right circumstances, even very small amounts of flammable gases or liquids can ignite and cause an explosion. This has led some gas-can manufacturers to equip their gas cans with flame arrestors.[7] These allow gases and liquids to pass through the device but stop or limit the spread of sparks and flames to help prevent a large fire or explosion.

The wire mesh in an inland tanker's gas pipeline acts a flame arrestor, limiting the spread of fire and preventing explosions.

Add conspicuous warnings

Principle
On-product warnings should communicate safety-critical information in an attention-grabbing and clear manner so that users can avoid hazardous situations.

Aren't warnings useless?
Warnings alert people to hazards—things that can cause harm. Hazards include moving gears, a hot surface, the shallow depth of a pool that is unsuitable for diving, a spinning propeller, or an alligator. Here, we will address the value of on-product warnings—a subset of the many different types of warnings people encounter in society.

First, dispense with the notion that warnings are useless or that everyone ignores them. Clearly, some warnings are at least somewhat effective at changing the behavior of some people some of the time. Is this a reasonable statement? We think so. The point is that warnings can, and often do, communicate important information to people so that people have the choice to take precaution. The key is to present the information in an attention-getting, clear manner that ensures people will understand the hazard, the potential harm that could arise due to exposure to the hazard, and the recommended way to avoid such exposure.

Effective warning labels should be:

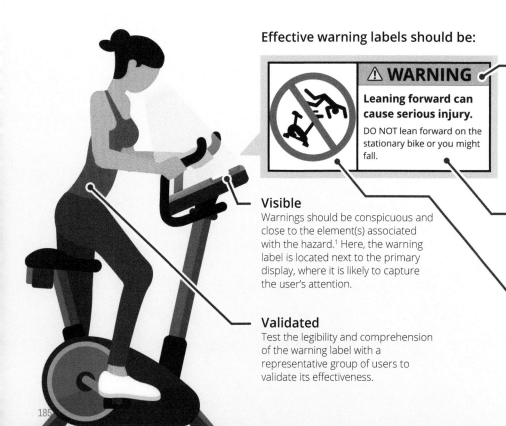

Attention-grabbing
Warnings should be sized appropriately, feature attention-grabbing graphics, and utilize color-coding and signal words prescribed by the ANSI Z535 series of standards. In this case, the orange panel and "WARNING" signal word are appropriate because of the potential for serious injury or death (e.g., if the rider hit her head).

Visible
Warnings should be conspicuous and close to the element(s) associated with the hazard.[1] Here, the warning label is located next to the primary display, where it is likely to capture the user's attention.

Validated
Test the legibility and comprehension of the warning label with a representative group of users to validate its effectiveness.

Clear
Use active voice, be concise, and "use definite terms and graphics commonly associated with the hazard."[2]

Visual
Use simple pictograms that visually reinforce the hazard, the actions necessary to avoid the hazard, or the potential harm. In this case, the graphic reinforces that the user could be violently thrown from the stationary bike.

Anatomy of a warning label

Decades ago, the American National Standards Institute, Inc. (ANSI) published the Z535 series—a series of standards for safety signs and colors.[3] These standards have guided many designers in the US to employ reasonably consistent warning information presentation styles. Common warning label elements include a signal word, an illustration, and a concise description of (1) the hazard, (2) the potential harm, and (3) the means of avoidance. These elements may be placed in many possible formats designed to facilitate readability. A warning might be prohibitive, which means it directs people to avoid doing something. "Keep hands away from the blade" is one such prohibitive message that describes the potential harmful behavior to avoid. Messages can also be permissive and instead describe the safe, hazard-avoidance behavior. For example, "Wear eye protection."[1]

Safety symbol panel[1]
Visually reinforces the hazard, the actions necessary to avoid the hazard, or the potential harm.

Signal word panel[4]
The color and signal word indicate the probability and severity of the harm if the warning is not heeded.

Message panel:[1]

1. **Hazard description**
   Alerts user to the hazard. In this case, the hazard is the airbag.
2. **Consequence statement**
   The potential harm that will or might occur if the hazard is not avoided. In this case, the consequence is death or serious injury.
3. **Action statement**
   Guidance for avoiding the hazard. In this case, the suggested action is to not place the child on the seat in front of the airbag.

Potential outcome	Color-coding and signal word
Will result in death or serious injury.	⚠ **DANGER**
Could result in death or serious injury.	⚠ **WARNING**
Could result in minor or moderate injury.	⚠ **CAUTION**
Important information but not injury-related (e.g., property damage).	*NOTICE*

Are warning labels really necessary?

In the case of a product liability lawsuit, a product manufacturer must be able to demonstrate that it applied state-of-the art design practices when developing and validating on-product warnings. Failure to adequately warn product users about hazards has led to many results in favor of the plaintiffs in such lawsuits.

Much of the guidance in this chapter comes from the ANSI Z535 series of standards, which includes:[5]

- ANSI/NEMA Z535.1-2006 (R2011), Safety colors
- ANSI/NEMA Z535.3-2011, Criteria for safety symbols
- ANSI/NEMA Z535.4-2011, Product safety signs and labels
- ANSI/NEMA Z535-2011, Safety color chart

Prevent scalding

Principle

For products that contain or produce hot fluids or steam, design them so that they are not susceptible to spilling or leaking, or so that they keep fluid temperature below a certain level to reduce the likelihood of scalding the user.

What is "scalding?"

There are many, perhaps even endless, numbers of ways to be burned accidentally, and many of them involve moisture in the form of hot fluid or steam. When the cause involves hot liquids or steam, it is called "scalding" and can lead to first-, second-, or third-degree burns.

Preventing scalding in an industrial environment is a big task, often requiring careful engineering of piping, valves, and other components to avoid accidental and forceful releases of hot water and steam. Protection is also afforded by wrapping pipes in insulating material that forms an initial barrier to hot contents reaching workers.

One one thousand, two one thousand

Producing a serious burn can be a matter of seconds.[1]

Water temperature	Time to produce 3rd-degree burn
155°F	About 1 second
148°F	About 2 seconds
140°F	About 5 seconds
133°F	About 15 seconds
127°F	About 1 minute
124°F	About 3 minutes
120°F	About 5 minutes
100°F	N/A – safe for bathing

Showers: for cleaning, not scalding

In homes and public settings, the risk of being scalded is often related to handling hot water. It used to be more common for people to scald themselves with hot water from a shower, often when people flushed a toilet and the pressure in the cold water piping of the home's water system suddenly dropped. The pressure drop would enable more hot water to flow because the hot water was still under high pressure. Today, this risk is often addressed by setting water heaters to a safer temperature; safety organizations recommend setting a water heater to no more than 120°F, yet not all homeowners remain compliant.[2] This risk is also addressed by installing temperature control mixing valves upstream from a faucet or showerhead, some of which have temperature setting dials.

Designing kettles to maximize safety

A hot water kettle's design can reduce or increase the chance of scalding. Electric kettles, which are becoming increasingly popular, should contain some of the following features to maximize safety:

A **non-spill spout** prevents hot water from dripping over the spout and down the side of the kettle, which could potentially scald a user's hand.

A **secure, non-slip, cool handle** can prevent a user from picking the kettle up and dropping it (and its hot contents) due to the handle's temperature or a slippery grip (see *Principle 40 – Make it slip resistant*).

A **properly-placed handle** enables users to maintain good control over the water-filled kettle based on the handle's position relative to the kettle's center of gravity.

A **cordless kettle**, wherein the kettle sits on a base connected to a power source, eliminates the potential for a child to pull on the cord, for example, and cause a kettle filled with hot water to topple over.

A **stable base** helps ensure that the kettle is not vulnerable to falling over if it is bumped in a busy kitchen, for example.

CAUTION: Contents may be HOT!

Hot drink spills are common culprits for scalding. A seemingly fictional but actual lawsuit was brought by one individual who spilled hot coffee on herself and was initially awarded $3M in damages. Reportedly, the plaintiff was holding the coffee cup between her knees and removed the lid to add cream and sugar, causing the cup to tip over and scald her. Ultimately, the case was about the coffee's (claimed) dangerously high temperature (180-190°F), and less about how the customer handled the cup.[3] The lesson here is that hot beverages should be served at a satisfyingly high temperature, but not so high that it would scald someone. But, some people might not be convinced: one study involving 300 participants suggests that people prefer to drink coffee at a temperature of about 140 +/- 15°F,[4] a temperature that can lead to a 3rd-degree burn in just five seconds.

Why we watch over children (and why you should, too!)

According to Shriner's Hospital for Children, hundreds of children with scald burns are admitted to the emergency room every day. In fact, 90% of burn injuries to children ages 5 and younger can be attributed to scalds and other contact burns.[5] As such, ensure that children are kept safe from hot fluids or steam, such as hot drinks, hot baths or showers, or hot fluid on a kitchen stove.

Exemplar 8
Steam iron

Historically, irons have caused an untold number of skin burns, electric shocks, and household fires. The good news is that modern versions have many safety features that help keep the focus on eliminating wrinkles.

Model photo: Colourbox.com

Manage and stow cords, cables, and tubes
Principle 71 - pg. 169

Recoiling mechanism automatically retracts the power cable inside the iron, helping to protect the cable from damage while also eliminating a potential tripping hazard.

Protect against electric shock
Principle 75 - pg. 177

Double-insulated cable protects users from electrical shock that could be caused by an exposed, live wire. The power cable also has a strain-relief feature that protects against wire damage due to kinking.

Make it slip resistant

Principle 40 - pg. 99

Textured, non-slip grip helps prevent users from dropping the hot appliance during use, which could lead to burned hands and feet, among other types of injuries.

Shut off automatically

Principle 73 - pg. 173

Automatically shutting-off after several minutes of inactivity helps reduce accidents that could result from a forgotten iron.

⚠ CAUTION: Hot surface

Prevent scalding

Principle 80 - pg. 187

Stable base prevents the appliance from tipping over if a user bumps the iron (or the ironing board) during or after use.

Add conspicuous warnings

Principle 79 - pg. 185

Conspicuous warning label informs users about potential hazards and safe ironing practices.

Shield or isolate from heat

Principle 52 - pg. 127

Heat-resistant plastic casing prevents "touch-points," such as the handle, dial, and steam release button, from becoming hot.

Enable safety feature testing

Principle

If a product serves an essential safety function, enable users to test the product—or better yet, design the product to run automatic tests—to ensure the product is functioning properly.

The waiting game

Many protective products are in standby for most of their lives—quiet sentries looking for danger in all the right places, but almost never seeing it because of its rarity. Consider the carbon monoxide detector, close cousin of the smoke detector. Although a smoke detector might get the occasional opportunity to perform, such as when bread is burning in a toaster, a carbon monoxide detector is arguably unlikely to ever go off. This is because common household events do not typically lead to high carbon monoxide concentrations that trigger a carbon monoxide detector's alarm.

I smell burnt toast—my chance to shine!

Ground-fault circuit interrupter (GFCI) receptacles—devices installed near sinks and other potentially wet places to protect people from electric shocks (see *Principle 75 – Protect against electric shock*)—are another example of a product that might rarely need to perform its primary function. So are automobile airbag systems, automated external defibrillators (AEDs), and inflatable aircraft escape chutes. It is good to know that the safeguards will work when it matters most.

Where's the "Reset airbag" button?

Some safety features cannot be fully tested without ordeal, expense, creating another kind of hazard, or "wasting" a single-use feature. That is why vehicles do not feature an airbag inflation test. The test would not only damage the steering wheel cover or dashboard but would also render the airbag useless. Therefore, people must trust that airbags will work as intended when needed and that any potential failures will be detected by the vehicle's electronic diagnostic system—which will then alert users of a potential fault by displaying an airbag light (see icon on right) on the car's dashboard.

Testing safety features

Ideally, products will alert users to safety feature failures in an automated manner, but this is not always possible. Although a smoke detector will chirp to indicate it has passed its expiration date or is running low on battery power, some products require users to actively test the readiness of its safety features, typically by pressing a button. This is the case with the aforementioned GFCI receptacles. With a press of the TEST button, the product switches internal electrical contacts and triggers the protective mechanism—a breaker. This function reassures users that they are protected and, otherwise, will indicate a device failure. After a press of the TEST button, some GFCI receptacles use an LED to indicate whether the system is powered and working correctly. However, newer regulations require that GFCIs manufactured after June 28, 2015 feature an automatic self-test function[1]—an arguably better solution that does not rely on users to remember to test the system.

Automatic self-tests

As mentioned above, it is helpful for products to run automatic self-tests. Modern automobiles do this almost continuously and will alert the driver if a safety failure is detected, such as the loss of the anti-lock brake function. Other systems will perform what is referred to as a Power-On Self-Test (POST). Systems that perform a POST run self-tests only on startup or will produce audible and visual signals on startup so users can verify that the visual and audible alarm indicators are functioning prior to using the product.[2]

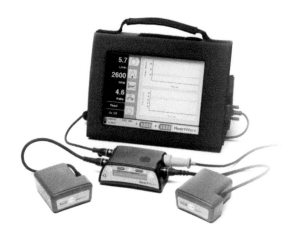

Some LVAD controllers (bottom center) run self-tests to ensure they are functioning properly.

As an example, let's consider Left Ventricular Assist Devices (LVADs)—mechanical pumps that are implanted in a person's chest to help pump blood through the body in case of heart failure.[3] In some LVADs, the pump's controller performs an automatic self-test when it powers on to ensure the controller is operating properly and has sufficient battery life. This self-test alerts users if there is an issue so they can replace the controller with a backup or connect to AC power.[4]

For safety-related products without an automatic self-test feature, users should be reminded to test the product intermittently to confirm that it is working properly. Otherwise, users are likely to forget to do so, unless they keep a calendar or set reminders via an electronic calendar (see *Principle 15 – Provide reminders*).

Products that do not run automatic self-tests should remind users to test the safety feature so that users do not have to remember to do so independently.

Add shape-coding

Principle
Product components should be shaped to reflect their function and to prevent misconnection or misuse.

Getting in shape

You can make a control a familiar shape to produce a useful association. Many examples of shape-coding are found in airplane cockpits. The tip of one lever is shaped like a wheel to create a strong association with its intended purpose—to raise and lower the landing gear. The tip of another control is shaped like a flap to create an association with raising and lowering the wing flap.[1] This shape-coding can help a pilot select the correct control by means of visual inspection, feel, or both.

The landing gear control in an airplane cockpit is shape coded to match the airplane's wheels. Photo (left): Colourbox.com.

Giving connecting parts a distinctive shape can also prevent misconnections. For example, you can shape various male and female connectors to ensure the correct ones match and that potentially dangerous—or at least inconvenient—misconnections cannot occur. That is one reason, among many others, that 120V and 220V electrical plugs and receptacles have different shapes and sizes. Another example is a patient monitor that has differently shaped ports for each cable that can connect to it.

Directional shapes can communicate purpose as well. An "Up" button in the form of an upward-pointing arrow, as opposed to an arrow graphic on a square button, just might help someone avoid pressing the "Down" button by mistake, such as when operating a forklift.

Many patient monitor cables have distinct shapes to ensure that users attach them to the correct port.

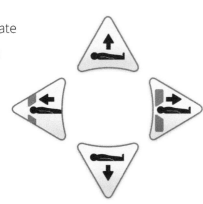

Shape-coding used on a CT scanner control panel.

What can go wrong when products aren't shape coded?

If a tube and its associated connector port lack unique shape-coding, there is potential for someone to connect mismatching components. For example, there have been instances in which a hospital worker has incorrectly connected a feeding solution tube to a patient's airway tube, killing the patient.[2]

Pen-injectors are used by laypersons and clinicians alike to deliver all kinds of medications, including insulin and growth hormones. Sometimes, it is not entirely clear to users which end contains the needle. This has led some people to inject their thumb, rather than their own or the patient's abdomen or thigh, for example. The result has been infusion of a drug at the wrong site or delivery of a drug to the wrong person (i.e., the person delivering care to a patient). Perhaps if the shape-coding were stronger, the needle end and the button end might be more visually distinct and recognizable.

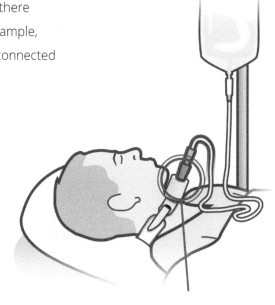

Due to a lack of shape-coding, nurses have connected feeding tubes to tracheostomy tubes, causing fluid to be delivered to patients' lungs.

Tips and tricks for effective shape-coded controls

Ensure that controls' shapes reflect their functions to strengthen the association between a control and its intended purpose.

When appropriate, combine shape-coding with other coding mechanisms (e.g., size- and color-coding) to make controls more conspicuous or distinct.

Ensure that cables and tubes can only connect to their matching ports. Make shape-coding overt, rather than subtle, to help users quickly identify and orient matching components.

Ensure that shape-coding is applied consistently to help users quickly identify the correct component or control and prevent potential misconnections and misinterpretations.

Alphonse Chapanis (1917-2002)
Alphonse Chapanis is considered to be one of the pioneers of human factors engineering. He combined visual and perceptual psychology with design engineering.

In 1942, Chapanis joined the Army Air Force Aero Medical Lab in Dayton, Ohio and researched user interface problems with air force equipment.[3] He discovered that many crashes were the result of pilots mistaking the nearly identical landing gear and wing flap controls. Chapanis used shape-coding to make the controls more distinct and, as a result, greatly reduced the frequency of B-17 bomber crashes during WWII.[1]

Provide backup power

Principle
Products that serve a critical function should contain a backup power supply to ensure continuous operation in the event of a power outage.

Why do we need backup power?

At well-equipped hospitals, a loss of electrical power from the external grid triggers a backup power generator that restores power quickly and might provide power for several days. In such cases, critical equipment (e.g., anesthesia machines, ventilators) used in operating rooms and intensive care units will continue working as usual. Sensibly, a product used in critical medical procedures will incorporate an uninterruptable power supply (UPS)—a battery or group of batteries with sufficient capacity to power the equipment as long as might be required. A UPS can take over electrical supply duties in a matter of milliseconds after its internal circuit detects a loss of voltage from the main power source.[1] The UPS covers for the time it might take for a generator to start up and for longer periods in the event that it does not.

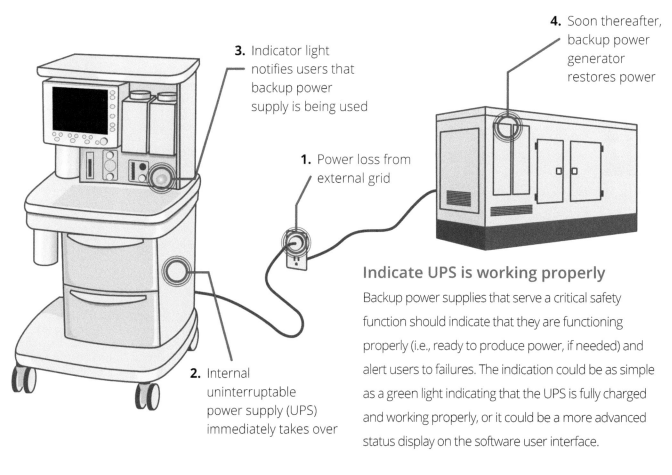

3. Indicator light notifies users that backup power supply is being used

4. Soon thereafter, backup power generator restores power

1. Power loss from external grid

2. Internal uninterruptable power supply (UPS) immediately takes over

Indicate UPS is working properly

Backup power supplies that serve a critical safety function should indicate that they are functioning properly (i.e., ready to produce power, if needed) and alert users to failures. The indication could be as simple as a green light indicating that the UPS is fully charged and working properly, or it could be a more advanced status display on the software user interface.

Different types of backup power supplies

Many systems that we interact with on a daily basis, including servers, elevators, telecommunication equipment, and lighting systems, depend on a backup generator or UPS. A few other examples include:

Ski lifts

Some ski lifts have both backup electrical power and backup motors to deal with temporary outages. In other cases, power can be provided by large, portable backup units.

Aircrafts

In an aircraft that loses main engine power, backups might come in several forms, including batteries, an auxiliary power unit (usually found in the plane's tail section), and a Ram air turbine, which is basically a high-tech version of a windmill. Backup power sources can be necessary in situations such as during bird strikes or when the engine fails due to ingesting volcanic ash.[2,3]

Sailboats

The wind and sail combined are a sailboat's power source. But, as every sailor knows, a sailboat is going nowhere in calm conditions. In such cases, a gas-powered motor is a handy and sometimes lifesaving backup feature.

Spacecrafts

The space shuttle (now retired) and other spacecrafts are well known for their many backup systems. Hydraulic power systems, which move flight control surfaces, among having other safety-critical purposes, have redundant and independent backup systems to account for primary system failures.[4]

Don't depend solely on color

Principle
Use redundant coding (e.g., texture, size, shape) rather than relying solely on color-coding, which might be ineffective in dimly lit conditions and for color-blind users.

Communicating through color
Traffic lights use color—green, yellow, and red—to communicate meaning. Yes, this is hardly a profound insight! But, the main point is ahead.

Traffic lights feature a limited number of color associations, which is likely why most people have an easy time interpreting them. In fact, the traffic light color meanings are so ingrained from a young age that few people will be consciously aware of the abstract association they're making between the colors and their associated meanings. A red circle means stop because we have learned the meaning of the colored symbol,[1] not because it is representational, as in the case of a raised hand symbol, for example. However, a small but significant fraction of people have some form of color vision impairment; approximately 1 in 12 men (8%) and 1 in 200 women (0.5%) with Northern European ancestry are color-blind.[2] People with color vision impairments do not see traffic lights the same way others with normal vision do. Instead, they must focus on the illuminated light's position,[3] learning that red is often on top, with yellow in the middle, and green on the bottom.

What color is this?
There are multiple types of color blindness. Some examples are:

Deuteranomaly (green weakness)
This is the most common form of color blindness. People with this condition have reduced sensitivity to green light and have difficulty distinguishing between red, orange, yellow, and green.[4]

Tritanopia (blue blindness)
This is a very rare form of color blindness, found in approximately 1 out of 10,000 individuals. People with tritanopia have difficulty distinguishing yellow from violet and green from blue.[5]

Monochromacy (complete color blindness)
People with monochromacy don't see any colors; they only see different shades of gray. As such, normal daily activities can be very difficult for them.[5]

Enhanced communication through position, shape, and size

The traffic light example on the previous page exemplifies how using more than one code—color and position—enables color-blind people to interpret traffic signals. As shown in the illustration on the right, adding other, and arguably stronger, codes such as shape and size could make a traffic light even easier to interpret for color-blind drivers. Notably, this stronger coding would also benefit drivers with normal color vision because the three signals are more distinct. Alas, the conventional traffic light is probably here for the long haul.

When it comes to communicating safety-critical information, more than one communication method can be essential. That's why it's sometimes best for warnings to be annunciated visually and audibly to accommodate people with vision and/or hearing impairments. Sometimes, triple-coding is even possible without really complicating the design. For example, a wearable device might produce audible and visual alarms along with a vibration.

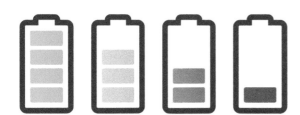

Example: Battery power
Correctly interpreting a battery or oxygen tank level can be safety-critical. As such, it makes sense to use more than just color to indicate the amount remaining. In the example above, the battery level is indicated by both a change in color and the number of bars.

Example: Gas adjustment knobs
Anesthesia machines have color-coded knobs, hoses, connectors, and pressure gauges.[6] Anesthesia machine knobs also have different textures that help users identify and differentiate between the knobs by touch.[7]

Are you color blind?

Some people can go years without knowing they are color blind because the effects of color blindness seem relatively minor. The most well-known color vision deficiency test is the Ishihara color blindness test,[8] which consists of 38 plates with dots of different sizes and colors. What do you see in the image on the left? Most people will see the number "74," but color-blind people might see "21" or no number at all. To complete the Ishihara color blindness test yourself, visit http://www.color-blindness.com/ishihara-38-plates-cvd-test/.

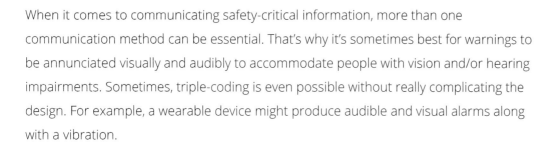

Today, there are special Enchroma glasses that enable people with red-green color blindness to better distinguish red and green. Visit the Enchroma website (http://www.enchroma.com) to see the emotional reactions of people wearing the glasses and seeing "real" colors for the first time.

Did you know Facebook is supposedly blue because Mark Zuckerberg is red-green color blind?[9]

Enable emergency calls

Principle
Certain products should enable users to call for help in an emergency, or they should do so automatically.

"Help. I've fallen and I can't get up!"

This statement, popularized in a commercial for an emergency communication system,[1] has since become a running joke, at least in the US. But, the need to help someone who has fallen is a serious one. Products, such as emergency call pendants that people can wear around their neck, offer a modicum of personal security (see *Principle 39 - Prevent falls*). The products work by sending a signal to a base station that can then communicate with a central facility to summon help to the stricken individual.

Firefighters are usually rescuing others, but firefighters might also require rescue in the event they are caught in a backdraft or building collapse. That is why many firefighters use portable radios equipped with an emergency button they can push when they cannot speak or are incapacitated in some manner.[2] Some prison guards carry radios with the same capability.

From 1988 to 2000, many highways were equipped with emergency call boxes used to report accidents and seek other emergency services.[3] The boxes have largely been phased out by the proliferation of cellphones. Interestingly, cellphones themselves have an emergency call feature. They can be used to call the local emergency number (e.g., 9-1-1 in the US) without having to first unlock the phone by entering a password.

"Mayday! Mayday! Mayday!" is the conventional expression to report an emergency over a radio. In 1923, it is widely believed that the senior radio officer, Frederick Mockford, working at the Croydon Airport in London, was asked to come up with an international distress call. He came up with "Mayday," which sounds like the French "m'aider," which is short for "venez m'aider"—in English this means "come and help me."[4]

Personal locater beacons

Another type of emergency call device is the personal locater beacon. The avalanche RECCO® system is often used to find people buried in snow. The system consists of a reflector, attached or sewn into people's winter gear, and a receiver used by the rescue teams. Rescuers will send out a search signal, which is then echoed by the reflector, directing the rescuers to the victim wearing the reflector.

Another type of personal locater beacon is integrated into the high-end Breitling Emergency watch. The watch is equipped with a dual-frequency distress beacon that guides rescue parties to its location, where presumably one would find the wearer in need of help. The user must unscrew two protruding crowns and pull two antennas out of the watch to activate both 406 MHz and 121 MHz rescue signals.[5] The signal is detected by a satellite that relays the message to the Mission Control Center (MCC). There, the message is decoded and sent to the nearest Rescue Coordination Center (RCC) so a rescue team can be dispatched.[6]

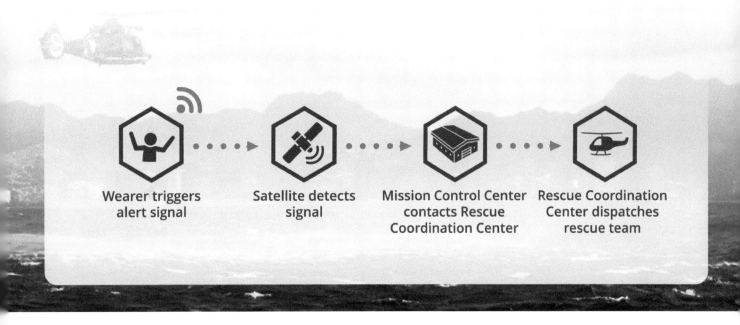

Wearer triggers alert signal → **Satellite detects signal** → **Mission Control Center contacts Rescue Coordination Center** → **Rescue Coordination Center dispatches rescue team**

Characteristics of emergency call buttons

The challenge is to ensure that emergency call functions can be activated easily in an emergency but not actuated unintentionally, causing false alarms. Emergency call buttons should:

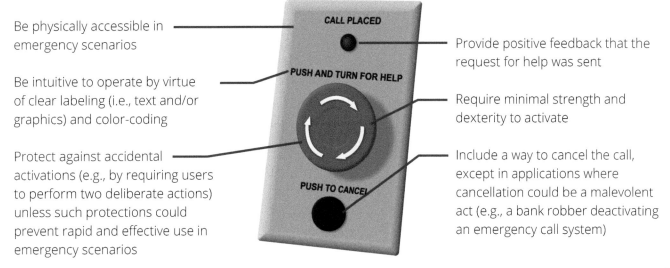

Be physically accessible in emergency scenarios

Be intuitive to operate by virtue of clear labeling (i.e., text and/or graphics) and color-coding

Protect against accidental activations (e.g., by requiring users to perform two deliberate actions) unless such protections could prevent rapid and effective use in emergency scenarios

Provide positive feedback that the request for help was sent

Require minimal strength and dexterity to activate

Include a way to cancel the call, except in applications where cancellation could be a malevolent act (e.g., a bank robber deactivating an emergency call system)

Use graphical instructions

Principle
Use graphics to make instructions and warnings more clear, attention-getting, and inviting, thereby increasing the likelihood that users will comprehend and follow them.

Graphic styles
Some people are disinclined to read narrative instructions and warnings, but they might pay attention to instructions that mostly use graphics instead of text. That is because graphics can be attention-getting and communicate messages more quickly than the equivalent text.[1] Some common graphic styles include:

- **Simple icon:** Simple lines and shapes create a symbolic representation of an actual object or abstract concept. This style is best used to facilitate rapid identification of a control or type of information (e.g., a question mark for help information).

- **Line drawing:** Lines and solid black and white fills create a simple image of an object. This style is best used for simple devices and when printing in color is not an option.

- **Illustration:** A mix of lines, solid fills and gradients create a representative image of an object. This style is particularly effective because images have depth and accurately depict colored components while having a simple appearance. This style is best used when printing in color is an option.

- **Photorealistic rendering:** Detailed illustrations that look as realistic as a photograph might be best used in marketing rather than instructional purposes. Photorealistic graphics tend to be too detailed to highlight specific graphical instructions.

Low graphic-to-text ratio *High graphic-to-text ratio*

Graphic-to-text ratio
Strive for a high graphic-to-text ratio (i.e., more graphics than text) in instructional documents. A text-heavy instructions for use (IFU) document is likely to dissuade users from reading the IFU because it looks intimidating and time-consuming to read. In contrast, users will likely perceive an IFU with a high graphic-to-text ratio (i.e., more graphics than text) and more blank space to be more welcoming and user-friendly, thereby increasing the likelihood that they read the instructions.

Which graphic style should I use?

Some people consider illustrations (i.e., somewhere between a line drawing and a photo realistic rendering) to be the best choice for instructional media. Effective graphics tend to be simple-looking illustrations that emphasize only the important details. Such drawings may be complemented by a limited amount of text and, perhaps, symbols indicating motion (e.g., arrows). The graphic must be large enough to enable viewers to recognize key details and should include additional visual elements that communicate essential meaning, such as wavy heat lines indicating a thermal hazard.

Bold and simple graphics can often communicate at a great distance, as evidenced by many traffic signs.

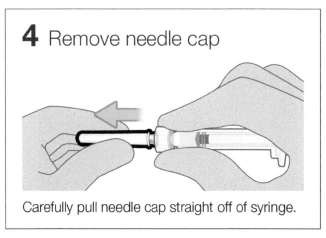

4 Remove needle cap

Carefully pull needle cap straight off of syringe.

Example of a graphical instruction that might appear in the Instructions for Use for a pre-filled syringe. The graphic emphasizes the cap and includes an arrow that communicates users should pull the cap to remove it from it from the syringe.

Did you know?

Evidence suggests a photo *really is* worth a thousand words.

 MIT researchers found that the human brain can process entire images in as little as 13 milliseconds.[2]

 We are more likely to remember images than words. This is known as the picture superiority effect.[3]

 Images are more attention-getting than words. Humans are wired to respond to visual stimuli.[4]

 People following directions containing both text and illustrations do 323% better than those following directions containing only text.[5]

And if I don't use graphics?

We have already established that instructions and warnings without graphics are prone to being overlooked. And, as one might imagine, overlooking critical instructions and warnings can have serious consequences.

For example, overlooking (or not comprehending) a written warning might lead a curious child to open an electrical utility box and receive an electric shock from an energy-storing transformer.

However, a clear graphic might alert an appropriately cautious child to leave the utility box unopened. In fact, good warning design suggests warnings should include attention-getting graphics that visualize the hazard and/or potential harm (see *Principle 79 – Add conspicuous warnings*).

Account for untrained use

Principle
Products calling for training are still likely to be used by untrained users and, therefore, should still feature design-based risk mitigations that protect untrained and trained users alike.

Training can be a tenuous proposition

Certain products feature labels advising use by trained individuals only. It is easy to understand the motivation behind the prerequisite; trained individuals are conceivably more likely to interact with a product safely and effectively. Unfortunately, that prerequisite will not always be met, and untrained users might not understand the risks of proceeding untrained. Sometimes, people overestimate their ability to use a product without training, preferring a trial-and-error approach to learning, hoping for the best, or perhaps being oblivious to the risks they are taking. Sometimes, people are pressed to use a product due to difficult circumstances, or they are put "on the spot." And sometimes, training might not be available anymore, like for an older model product.

When safety is part of the job

Licensing and certifications can arguably be an effective way to ensure that someone in a specific role has the appropriate level of skill. For example, "It is an OSHA requirement to train anyone who operates a powered industrial truck. This requirement applies whether they operate the truck daily or once a year, regardless of the amount of their experience."[1] To fly passenger planes, pilots must not only earn an advanced license but must also be trained to operate a specific aircraft (e.g., Boeing 787, Airbus 350, Embraer 175).[2] Similarly, a perfusionist needs to be certified to operate specific heart-lung machines.[3] These are all reassuring examples of required training.

Unfortunately, job-specific training can be uneven. A registered nurse might be asked to work with a new, unfamiliar infusion pump, leading the nurse to misprogram the pump and deliver too much medication to a patient. Similarly, an anesthesiologist is not always required to demonstrate competence on the use of an anesthesia machine.[4] Apparently, there is a presumption of competence at using a potentially unfamiliar product.

Photo: Colourbox.com

Model photo: Colourbox.com

Consumers should read a user manual before using a product, but can you guess why they might not bother?

A
B
C

Possible answers:⁵

A: I get it, the control knobs are labeled like my stove, so it works the same way...and I'm hungry now!
B: I see my neighbor use a lawnmower all the time; it can't be that hard. C: I have a driver's license; I can drive any car.

To account for untrained use, consider a variety of tactics

Designers can try to make a product as simple and intuitive as possible to reduce the product's reliance on training. However, because it's not always practical or appropriate to simply remove complex features that might require training or skill to use properly, consider the following tactics that can potentially accommodate untrained users:

Test the design
User testing is an important step in the design process. In particular, you can check how intuitive a product is by conducting a usability test with untrained subjects.

Include additional user support
Include additional user support (e.g., quick reference guide, 24/7 help line) that can serve as a "lifeline" for users who are unsure how to perform a given task.

Grant progressive access
Provide access to simple features first, and then provide add-ons as the user gains experience or undergoes more training modules. For example, enable access to features based on locks/passwords linked to specific credentials and training.

Maximize intuitive use
Intuitiveness arises from designers applying good design principles, such as "universal design"⁶ and many more of the principles in this book, including clearly labeling displays and controls, using clear graphics, and using shape- and color-coding.

Control access
Some medical devices are sold through specialty pharmacies that require users to present proof that they have received training. Or, a trainer must sign off that training was given and that the user has passed a competency test.

Embed training
Embedded software training modules on devices can get users up-to-speed and potentially eliminate the need for in-person training, which can be prohibitively expensive or otherwise infeasible.

Give hints and prompts
Provide on-screen hints, tips, and coaching that appear (or are placed) in the context of a specific step or area of interaction so that guidance is immediately available and easy to find when needed.

Let users set the pace
Untrained and new users might need more time to conduct tasks than more experienced, expert users. Let users conduct steps at their own pace, and repeat steps as needed.

Provide guidelines

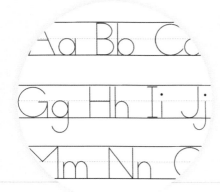

Principle
Provide literal guidelines on products and in public spaces to indicate the desired path that a person or machine should follow to remain safe, or to indicate a boundary line that people should not cross to avoid harm.

Let's set the record straight
The use of the term "guideline" has expanded to describe guidance or advice, such as the content in this book. But, technically speaking, a guideline is a line indicating a desired path or a boundary that should not be crossed.

Roadways have guidelines in the form of a single or double yellow line, a single dashed line, or a solid white edge marker. These guidelines help keep people driving in the proper lane to prevent collisions. There are countless more forms and places in which you might find guidelines.

One form is boundary lines that guide people to stand back from a hazard, such as opening bus doors or a moving forklift. Perhaps one of the most well-known boundary lines is the yellow, often-textured area on a train platform that people should remain behind to stay at a safe distance from an approaching train. Another form indicates the limits for storing or stacking things. For example, the fill line on a sharps container indicates when it is full and cannot be closed properly if more sharps are added. Similarly, the height limit indicated for stacking pallets in a warehouse is intended to prevent pallet towers from tipping over onto bystanders.

Guidelines aren't only for people
Automobiles equipped with lane assistance technology often use cameras to get a fix on lane markings. This makes it possible for the car to warn the driver of drift toward the lane's edge and, if needed, automatically perform corrective steering to stay in the lane. Some systems can actually keep the vehicle centered in the lane.

Examples of guidelines

Below are four examples of ways that guidelines can be used to ensure safe interactions:

Indicate safe zone

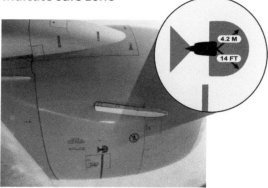

Jet engines have lines drawn on the side of their nacelles to indicate boundaries past which someone working around a running engine could be sucked into it. The lines are also supplemented with a diagram indicating the distance someone should remain away from the nacelle to avoid injury.[1]

Indicate limits

Redlines on tachometers indicate maximum revolutions per minute (RPM) limits beyond which potentially catastrophic engine damage could occur. Some mechanisms with redlines are also protected by governors that prevent over-speeding. Lacking such features, an engine could seize, potentially causing the vehicle to go out of control.

Indicate hazardous path

A miter saw or chop saw, as it is sometimes called, projects a laser line onto a plank of wood to indicate the blade's path. This offers the advantage of guiding accurate cuts and indicating the danger zone for hands and fingers.

Indicate projected location

An increasing number of cars have backup cameras that help people drive in reverse. Camera views are augmented by guidelines indicating where the car is headed based on its current position and steering wheel position. The augmented reality view helps people avoid objects, including people.

Use voice prompts

Principle

In cases were people are stressed or their focus is directed away from the device or its display, use voice prompts to provide rapid and clear audible guidance and feedback.

The power of voice prompts

Although a well-designed graphic can be more attention-getting and communicative than text, a verbal instruction or voice prompt can be even more commanding and clear. This is especially true in cases where people are stressed, preoccupied, or focused elsewhere than a device's screen. Accordingly, voice prompting should be considered for applications in which rapid and compelling communication is critical.

Automated external defibrillators (AEDs), which are used to rescue someone experiencing sudden cardiac arrest, are a particularly good application for verbal instructions. AEDs are typically located in public spaces, where the person using it is unlikely to have ever used it before or have any familiarity with how it works. Loud, clear, simple, and properly timed commands are what the user needs to perform the necessary actions to help the person in need.

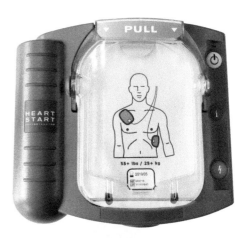

AEDs, like the one above, use voice prompts to guide the user through the high-pressure situation of applying CPR.

AEDs often provide verbal commands to keep users on track[1]

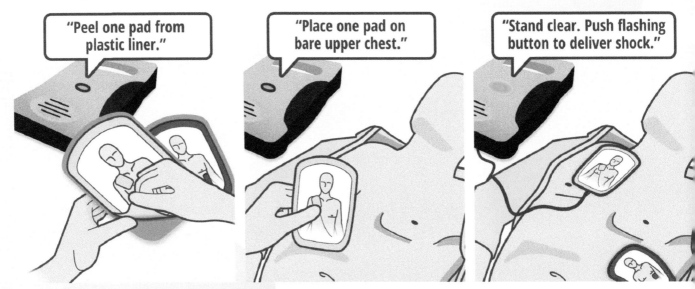

Sound the alarm!

Verbal commands are a powerful way to communicate to an airplane pilot that there is an imminent hazard, such as indicating terrain ahead and the need to "Pull up" (see *Principle 18 - Predict hazardous situations*). Such voice commands can break through what might be a pre-existing condition of visual sensory overload. A system called TCAS II (traffic collision avoidance system) instructs the pilot of an aircraft on a collision course with another one to perform evasive maneuvers without delay. Voice commands include "Climb, climb" and "Descend, descend." These commands, among others, are termed "resolution advisories."[2]

"Climb, climb"

"Descend, descend"

"...move to fresh air"

Some smoke and gas detectors issue verbal alerts and commands as well. When the Nest Protect[3] smoke and carbon monoxide detector detects carbon monoxide, a female voice states: "Emergency. There's carbon monoxide in the hallway. Move to fresh air." At the first detection of smoke, "she" will state: "Heads up. There's smoke in the kitchen. The alarm may sound."

Male versus female voices

When voice prompt systems were first introduced into various products, a female voice was considered most attention-getting, presumably because the higher-pitched voice of a typical female would be more penetrating than the lower-pitched voice of a typical male. But, recent research has suggested that there is no substantial difference in voice command effectiveness. As such, the choice of a male or female voice should be determined based on the nature of background noise with regard to the pitch of the speaker's voice and the potential for masking.[4] Masking occurs when a stimuli does not stand out well from similar sounding background noise.

Can you hear me?[5,6]

Below are some guidelines for designing effective voice prompts:

- **Use an appropriate delivery rate**
  Use about 170 words per minute to ensure words are distinguishable and the user has time to process the information as it is being delivered.

- **Use concise voice prompts**
  Use brief voice prompts that are easy to interpret and remember. Avoid ambiguity.

- **Use simple terminology**
  Ensure that voice prompts do not use technical details that are difficult to understand.

- **Update the user**
  Ensure that the system informs the user about what it is doing. For example, an AED annunciates "Shock delivered" once it has delivered a shock.

Enable fast action in an emergency

Principle

Make sure users can perform safety-critical tasks quickly and easily during emergencies.

Making a quick escape

An emergency stop control enables one kind of critical action: halting an action or process immediately (see *Principle 51- Enable emergency shutdown*). But, there are other emergency actions that might need to happen immediately.

Opening an airliner's exit door is one example. Aircraft manufacturers go to great lengths to make these doors relatively intuitive to open by limiting the number of required actions and providing instructions on both the door and in instructional cards. Instructions must be printed in languages that the crew speaks to the passengers—typically the most common languages of a flight's origin and destination.

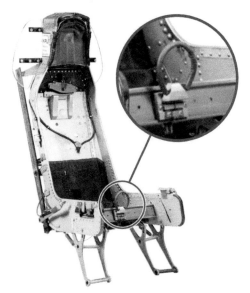

Similarly, jet cockpits are equipped with ejection seats that might need to perform their life-saving action in an instant. Some ejection seat handles are located between the pilot's legs, requiring the pilot to assume proper ejection posture before they can activate the ejection mechanism.

Triggering the ejection seat must require a deliberate rather than inadvertent action because an injection from a jet moving at the speed of sound can subject a pilot or crew member to a 20G force.[1] Plus, an ejection also results in the loss of an expensive aircraft and risk to property and persons on the ground.

A Saab 32 Lansen military aircraft ejection seat with a red ejection seat handle between the legs.

Easy access

Activating a reserve chute when the main chute fails is another example of a safety-critical task that needs to be performed quickly. The reserve chute release mechanism requires only a single action and is often color coded red to convey its emergency purpose. There are even devices available that use a pyrotechnic charge to automatically deploy the main or reserve parachute if the skydiver descends below a preset altitude—a nice safeguard for a skydiver who might be knocked unconsciousness, for example.[2]

Another example is the actuation of fire suppression systems in race cars. Drivers dealing with a fire need only press a button or pull a handle to trigger the onboard extinguishers. One thing to consider when enabling an emergency action is the possibility of accidental activations that could cause harm and property damage when there is not an actual emergency. In the case of a fire suppression system, it does not make sense to require a confirmation action because it might cause a tragic delay in extinguishing the fire. But, placing a cover over the quick action control might be a reasonable safeguard.

Fast action on software user interfaces

More and more, software user interfaces are replacing hardware user interfaces, and it's just as important to ensure users have fast access to critical software controls. Imagine a touchscreen-based hemodialysis machine that removes blood from a patient, filters it, and then returns it to the patient. A nurse needs to be able to intervene and stop the system immediately if there is a problem. So, requiring the nurse to access a menu and then select "Pause treatment" from a list of options does not make sense. A better solution would be to have a large "Pause treatment" button on the top-level screen to ensure it is always accessible in an emergency.

In some cases, hardware and software work together to provide access to critical controls. For example, to make an emergency call on some iPhones, users can rapidly press the power button five times to display an emergency SOS control. The feature enables users to surreptitiously call for help in dangerous situations and enables strangers to call for help without needing the phone's unlock code[3] (*see Principle 85 – Enable emergency calls*).

When designing a software user interface that requires fast access to emergency controls, consider these additional guidelines:

✔ Use contrasting colors (e.g., red), make the control large, and add a clear label (e.g., "STOP treatment") to make the emergency control immediately recognizable.

✔ Place critical displays and controls in a convenient, readily accessible location. Choose a location that minimizes potential for accidental contact, especially when designing for touchscreens.

✔ Allow for easy cancellation in the case of accidental actuation.

Exemplar 9
Automated external defibrillator

Automated external defibrillators (AEDs) analyze the heart's rhythm and deliver an electric shock to help the heart re-establish a normal rhythm. They also provide instructions to deliver cardiopulmonary resuscitation (CPR). Different AED models are intended for use by medical personnel, laypersons, or both.

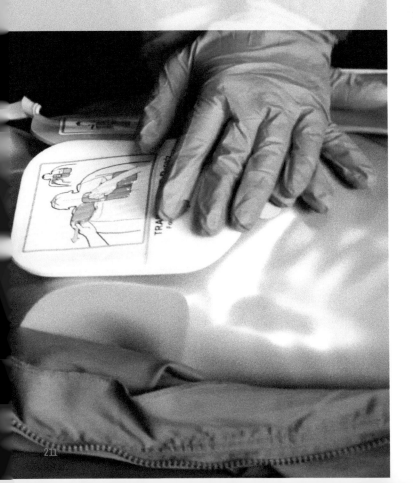

Make packages easy to open
Principle 65 - pg. 155

Electrode package has pull tabs that enable users to open it and access the electrodes quickly and easily.

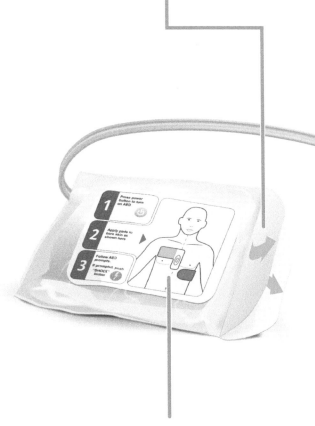

Don't depend solely on color
Principle 84 - pg. 197

In addition to being color coded, electrodes are also shape coded, enabling users to accurately associate them with those shown on the chest place diagram.

Enable fast action in an emergency

Principle 90 - pg. 209

AED is equipped with all necessary accessories, some of which are pre-attached before use, enabling users to perform an emergency defibrillation.

Use graphical instructions

Principle 86 - pg. 201

Simple, on-screen, graphical instructions guide users through every step, from placing the electrodes to delivering compressions.

Enable safety feature testing

Principle 81 - pg. 191

Self-check feature enables users to confirm the AED is charged and functioning properly (i.e., ready to deliver a shock and guide CPR) before use.

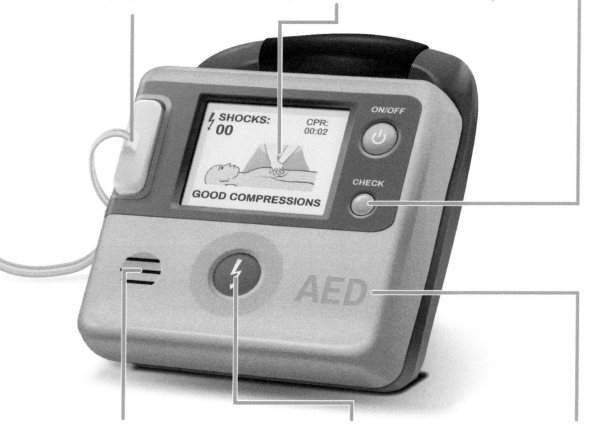

Use voice prompts / Account for untrained use

Principle 89 - pg. 207 / Principle 87 - pg. 203

Loud, clear, concise, and properly timed voice prompts guide users (who might have never used an AED before) to operate the AED correctly and safely.

Provide illumination

Principle 22 - pg. 61

Back-lit displays, indicators, and controls ensure the AED can be used in dimly lit environments. Controls light up when they become relevant to the task at hand. For example, the "shock" button flashes only when the AED is ready to deliver an electric shock.

Use sensors

Principle 56 - pg. 135

Sensors and associated software algorithms determine if the patient's heart rhythm is normal. If the AED detects an abnormal rhythm, the device advises the user to deliver an electric shock.

Make it touch free

Principle
In cases where touching a device could be hazardous (e.g., spread harmful bacteria), enable touch-free (i.e., hands-free) operation.

Avoiding that personal touch
A modern public restroom can be a touch-free tour de force. There's no need to touch a lever to flush the toilet, turn a faucet handle, push the plunger on a soap dispenser, or press a button to start the hand dryer. It's all handled (pun intended) without the need to use one's hands.

Like other modern conveniences, touch-free products are becoming more ubiquitous. Considering that many young people today have never dialed a rotary telephone or turned a lever to roll down a car window, might tomorrow's youth become so reliant on touch-free technology that they consider manually flushing a toilet or pushing a door open to be arcane actions?

While touch-free gadgets are convenient and might be considered a luxury, the safety-related benefit of touch-free operation is better hygiene. Given the number of bacteria that can reside on common touch surfaces, touch-free tech can help reduce the spread of germs. For example, in healthcare environments where acquired infections are a major concern, touch-free technology can reduce "contact transmissions."[1]

Then again, touch-free solutions are not a panacea. A toilet with a hands-free flush mechanism does not necessarily spare people from contamination by infectious bacteria unless the lid is down during the flush. Otherwise, a forceful flush can eject bacteria-containing mist into the air. Similarly, a hand (or jet) dryer can send germs flying off of hands toward other people. In fact, jet dryers can throw germs a distance of up to three meters![2]

"What do we do...use our hands?!"

Bacteria counts per square inch
based on a survey conducted by the Hygiene Council of 35 US homes[3]

13,227
Kitchen faucet handle

6,267
Bathroom faucet handle

295
Toilet seat

Touch-free tech is a growth industry

Designers can employ numerous ways to actuate touch-free devices. Some touch-free products use passive infrared to detect motion, actuating when they detect the presence of a warm object (as with paper towel dispensers) or the departure of one (as with toilet-flushing mechanisms). Some products radiate microwaves and look for a Doppler Effect to detect moving objects.

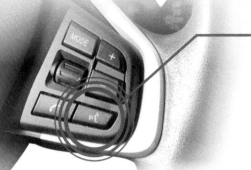

And, some products take advantage of voice-controlled technology, such as digital "assistants" like Apple's Siri and Amazon's Alexa. For example, voice-controlled automobile infotainment systems enable drivers to keep their hands on the wheel and eyes on the road.

Here are some more examples of touch-free applications that enhance safety:

 Automatic doors enable those with physical limitations to hold onto a mobility aid (e.g., a walker) rather than risk falling while trying to open a door.

 Automatically activated nighttime lights can help users stay on path and avoid colliding with or tripping over hazardous objects.

 Hands-free liftgates can be opened with a wave of the foot, relieving users with full hands from having to assume awkward positions that could lead to strain or injury from dropping objects.

 Automatic feeding mechanisms, such as those on paper shredders, keep the user's hands out of harm's way (see *Principle 21 - Incorporate automatic feeding mechanism*).

Safeguarding touch-free features

Despite its many potential benefits, touch-free tech is not a safety slam dunk. Use errors and product malfunctions can still occur and can interfere with the availability and effectiveness of touch-free features. Here are some possible risk mitigations:

- A manual actuation option that can be used if the touch-free function fails
- Backup power (AC power backup for battery power or vice versa) to avoid loss of function (see *Principle 83 - Provide backup power*)
- A way to disable touch-free functionality if the feature should not be used in certain scenarios
- Instructions on how best to use the touch-free function (e.g., a screen prompt such as "You can say things like...")
- On-product labels guiding users to properly interact with the touch-free product (e.g., an icon showing users where to place their hands to dispense soap)
- Enabling touch-free operation by only authorized individuals perhaps by responding only to certain voices, or requiring users to carry an RFID fob
- An activating commands (e.g., "Hey Siri") and directional or range-limited sensors that help prevent unintended activation

Guard against overinflation

Principle

Products that must be operated within a safe pressure range should warn or protect users against underinflation and overinflation.

Sharper means safer

Practically everybody has overinflated a balloon and caused it to burst. Some people even pop balloons on purpose. Overpressurization can be OK for party balloons but arguably not OK for products that can kill or maim if they burst.

Large truck tires, for example, will release tons of built-up pressure if they burst due to overinflation, causing "blast injuries." Injuries due to bursting objects include face lacerations, bone fractures, pneumothorax (collapsed lung), and brain edema, in addition to more grave injuries.[1]

An inflation cage guards against serious physical harm if a large tire explodes during inflation.

One solution to tire overinflation also addresses the problem of underinflation, which can also lead to adverse events such as automobile roll-over accidents. The solution is an automatic system that maintains tire pressure by increasing tire pressure up to a prescribed level (and no higher) and then releasing excess pressure created by a tire heating up (e.g., from warm weather and/or rolling friction).[2] Another solution is to install pressure relief valves that simply vent excess pressure whenever it develops.

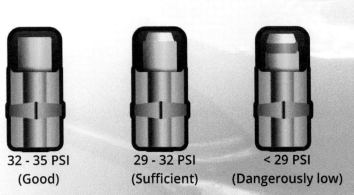

32 - 35 PSI
(Good)

29 - 32 PSI
(Sufficient)

< 29 PSI
(Dangerously low)

Some tire valve caps indicate the tire pressure to help drivers avoid underinflating their car's tires, thereby preventing premature tire wear and potentially dangerous blowouts.[3]

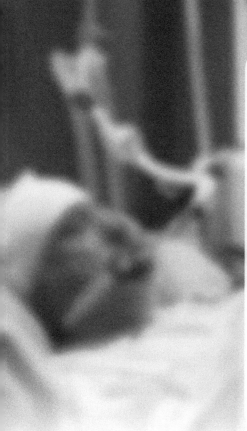

Adding inflation alarms to equipment

Products can also provide active pressure warnings. For example, consider ventilators, which are machines used to breathe on behalf of people who cannot. Forcing too much air into the lungs can cause excess pressure, resulting in acute lung injury and air leaks that deprive the patient of oxygen.[4] To address this risk, ventilators are equipped with pressure sensors and alarms. Similar protections are necessary when insufflating (i.e., inflating) the peritoneal cavity to create the space and visual access required to perform laparoendoscopic procedures. In the event of high pressure, the insufflator's pressure sensors trigger alarms, while pressure-relief valves reduce pressure if it exceeds a preset limit. Lacking such protection, the device could overinflate the peritoneal cavity, potentially causing an air embolism.[5]

Add labels or other overinflation protection

What other kinds of products pose a hazard when overinflated? Believe it or not, exercise balls! The US Consumer Product Safety Commission actually had to issue a recall on one type they deemed to be dangerous. "The fitness balls were determined to pose a risk from overinflation which was causing them to burst without warning during use, thereby tossing the exerciser to the ground."[6] Forty-seven people reported incidents of the balls bursting and causing bone fractures and bruises. The recall covered 3 million balls. The publicized fix involved revised instructions regarding proper inflation and a caution to measure the ball's size to prevent overinflation.[7] Notably, some exercise balls are actually able to deflate slowly if punctured and are advertised as anti-burst.

INFLATION/DEFLATION INSTRUCTIONS
⚠ WARNING
• Serious bodily injury may result if inflation and deflation instructions are not carefully followed. Exercise balls MAY BURST from the effects of improper handling, which could cause serious injury from a fall.

In summary, consider these actions as possible protections against overinflation:

• Enable the user to set pressure limits.

• Indicate safe inflation pressure (e.g., via a label or instruction manual).

• Indicate the current pressure of the inflated item, perhaps with an attached gauge.

• Physically guard the pressurized item.

• Automatically maintain the correct pressure.

• Provide an alert and then an alarm when pressure is getting high.

A warning label describes the risk of overinflating or underinflating the exercise ball.[8] (Photo: Colourbox.com)

Label toxic substances

Principle

If a product contains substances that are harmful to its users, it should have a label stating the hazard and what do in case of harmful exposure.

Why is labeling important?

The modern industrial world produces, stores, and transfers an enormous amount of toxic substances, and people must be aware of the presence of toxic substances to take necessary precautions. This is particularly true for the parents of young children and first responders. Parents need to understand and secure substances that could harm children who might touch, taste, or swallow the liquids and solids, unaware of the consequences (see *Principle 63 - Childproof hazardous items*). First responders need to know what they are dealing with in an industrial spill scenario, for example.

The Federal Hazardous Substances Act (FHSA) mandates that all hazardous household products have precautionary labeling on their containers. In addition to giving consumers information about what first aid steps to take in the event of exposure, the labeling also helps ensure consumers know how to safely store and use those products.[1] Perhaps the most iconic instance of toxic substance labels is the poison label found on some household products. The familiar skull and crossbones (also called the Jolly Roger) has traditionally been used to label poisonous substances found in homes (e.g., cleansers). Mr. Yuk™ was introduced more recently as less of an "attractive nuisance" and, arguably, more of a communicative symbol to dissuade young children from interacting with toxic substances (see *Principle 35 - Make design features congruent*). The idea is to place a repellent Mr. Yuk™ sticker on all dangerous substances containers (see the Mr. Yuk™ label on the right), including those that might seem benign but can cause chemical injuries to the skin (e.g., laundry detergent packets and chlorine-based cleansers).[2]

In 2011, more people in the US died due to accidental poisonings (36,280) than in automobile accidents (33,783), according to the Centers for Disease Control and Prevention (CDC).[3]

Requirements for good labeling

Toxic substance labels, which may be considered warnings of a sort (see *Principle 79 - Add conspicuous warnings*), should be prominent, clearly state the hazard, and state ways to avoid the hazard. Any incorporated graphics should be easy to interpret correctly at a glance.

Safety labels should be tested to confirm they will last a long time in their intended use environments, which might expose the label to extreme temperature, humidity, UV light, and chemicals.[4] Certain labels (e.g., placards) also have to pass strength tests.

When potential toxic substances cannot be identified by their labeling, first responders can use products, such as the Thermo Scientific™ FirstDefender™ RMX Handheld Chemical Identification Analyzer, to rapidly identify unknown substances, including toxic chemicals, explosives, chemical weapons, white powders, and more.[5]

Label design standards

Specific standards should be consulted to determine a label's contents. The standards below present two commonly used systems to indicate the hazards of toxic materials and the severity of the hazards.

NFPA 704

The National Fire Protection Association (NFPA) developed the "NFPA hazard diamond" to help emergency personnel identify risks associated with hazardous materials.[6] Each of the four diamonds provide specific hazard-related information that a given chemical might pose during a fire.

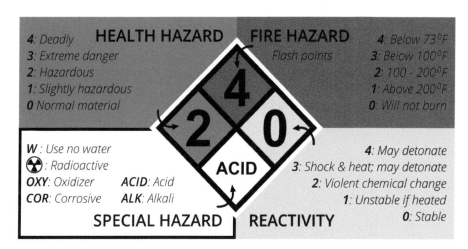

49 CFR 172.519

The 49 CFR 172.519 hazardous material standard provides guidelines for Department of Transportation (DOT) placards that inform transportation workers and emergency responders about material hazards:[7]

1. A symbol indicating the material's primary hazard class is displayed in the upper corner.

2. The hazard class name is displayed in the center.

3. The background color indicates the material's primary hazard class. For example, materials in the poisonous class are usually white, and materials in the flammable class are typically red.

4. A hazard class number (class 1-9) is displayed in the lower corner. Each hazardous material is assigned to a specific hazard class.

Include brakes

Principle

If a product must remain stationary or move slowly to be operated safely, ensure it has brakes that are visually conspicuous and physically accessible.

Hold up!

Safe equipment operation often depends on keeping the equipment in place, which, in turn, might require brakes. Some applications call for users to apply the brakes as needed to maintain a desired velocity or to secure an item in place. Other applications call for brakes to remain engaged until the user temporarily releases them, perhaps by squeezing a spring-loaded bar. Such a mechanism ensures the brakes automatically return to a safe, engaged state without further action from the user.

An unrestrained food cart can injure airplane passengers and cause cabin damage if it manages to get away from the flight attendant, perhaps due to turbulence or a sharp ascent or descent. In April 2016, a fully-stocked beverage cart on an American Airlines flight zoomed down the aisle during take-off and struck a passenger in the head, causing bleeding, loss of consciousness, and alleged "chronic traumatic brain injury and post-concussive syndrome."[1] To prevent such incidents, airplane carts are typically equipped with foot-applied brakes that flight attendants can easily engage and release. But, attendants still need to remember to do so at the right times.

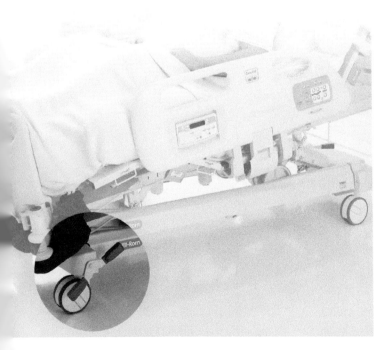

Modern hospital beds have wheels that enable patients to be transported without requiring patients to exit their beds. As important as it is for the beds to be mobile, it is equally important that they do not roll away once the patient has been transported to their destination. For example, some critical care beds are equipped with brightly colored brakes near the bottom-most left wheel. Some beds also provide audible or visual reminders when a brake is left disengaged (e.g., an alert displayed on a touchscreen that is embedded in the handrail).

Give me a brake

Additional examples of wheeled devices that utilize brakes to prevent unwanted movement include:

Airport luggage carts
Many airport luggage carts feature handles that also function as the braking mechanism. When the user releases the handle, the brakes engage automatically.

Wheelchairs
Wheelchair brakes are typically actuated by easily accessible levers located near the left or right wheel that seated users can push or pull.

Baby strollers
Most baby strollers have brakes located behind the rear wheels that users can quickly engage and disengage with their foot.

 Of course, brakes are standard equipment on moving vehicles, ranging from bicycles to cars to trucks to trains. Brake system evolution began in the 19th century with wooden block brakes used in horse-drawn vehicles and steam-driven automobiles.[2]

Design brakes so they are...

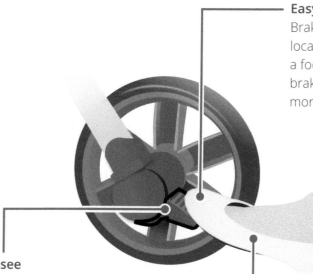

Easy to reach
Brakes should be easily accessible. They are often located near the affected wheel(s) so the user can use a foot to engage the brake swiftly. In some cases, a brake bar located close to the user's waist might be more accessible.

Easy to see
As with emergency shutdown controls (*see Principle 51 – Enable emergency shutdown*), it is important to make the brake mechanism visually conspicuous (e.g., by making the brake mechanism a bright color).

Easy to engage and disengage
There's no point making a brake easy to reach if it isn't also easy to engage. Many brake mechanisms require a single action—like a quick step of the foot—to engage and disengage the brake. It's also important to clearly convey the state of a brake (i.e., whether or not it is engaged).

Provide restraints

Principle

Personal restraints should be used to prevent and protect users when undesirable movement (e.g., falling from great heights, being ejected from a moving vehicle) could cause injury or loss of control.

Buckle up

Lots of products incorporate personal restraints, and many more products probably should. The obvious example of restraints serving life-saving and injury-prevention purposes is seat belts in cars. It is surprising that the US did not mandate automobiles to include seat belts as standard equipment for forward-facing seats until 1968,[1] even though they were patented 83 years earlier by American Edward J. Claghorn.[2] Even today, many school buses are not fitted with seat belts in the US because many states do not require them, which might be due in part to an arguably low death rate (about 6 children per year out of the 23.5 million children routinely riding school buses in a given year)[3]—a macabre fact that offers little solace to the families of the children killed.

Many other types of vehicles are or might be equipped with seat belts, such as airplanes, earth-moving equipment, and all-terrain vehicles. The benefit is clear—the restraints prevent people from being thrown from their seats due to sudden vehicle movements. Two-point belts are a baseline solution, three-point belts are a step up, and four-, five-, six- and seven-point belts (i.e., harnesses) generally offer the best protection, which explains their use in race cars, fighter jets, and child car seats. Specially designed restraints (i.e., harnesses) can prevent vehicle occupants from "submarining" in an accident. "Submarining" involves the lower body sliding forward and under the waist belt such that the belt no longer bears on the pelvic bones and cuts into the abdomen.[4] Some solutions further prevent "submarining" by incorporating a crotch belt.

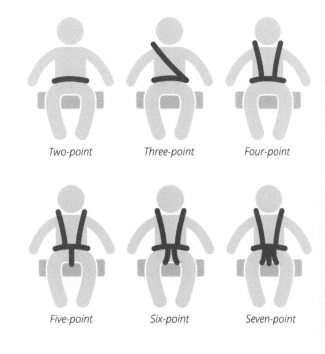

Two-point   Three-point   Four-point

Five-point   Six-point   Seven-point

 In 2010, over 190,000 of the 9.1 million passenger vehicle crashes involved the vehicle rolling over. Furthermore, 35% of all deaths from passenger vehicles involved a roll-over accident, resulting in more than 7,600 deaths. The majority of those killed (69%) were not wearing a seat belt.[5]

Different types of restraints

Whereas restraints such as seat belts usually take the form of strong fabric belts and metal connectors, restraints can take the form of more rigid (albeit padded) elements, such as those found on amusement park rides (see *Principle 26 - Incorporate a lockout mechanism*). Restraints are also used in several other applications, including the following:

Construction

Special harnesses are worn by brave individuals such as high-rise steel workers and window washers, who would otherwise be vulnerable to falls from great heights. When the first skyscrapers were built, it was said that for every $1 million spent on the building, one construction worker died.[6] Fortunately, the days of construction workers having lunch on top of skyscrapers and walking along steel beams positioned hundreds of feet in the air are gone, largely due to occupational safety laws.

Health care

Patient restraints are sometimes used in medical settings. For certain surgical procedures, patients are physically restrained to keep them in the best orientation for surgery and to prevent sudden body movements during the procedure. In mental health facilities, patients might be restrained to prevent them from hurting themselves or others. The FDA Guidance on Protective Restraints (1995) provides guidelines for medical protective restraint devices.

High-rise steel workers and window washers, who work at great heights, typically wear harnesses secured to anchorage points to prevent against fall injuries and/or death.[7]

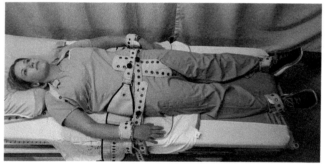

Some patients in mental health facilities are restrained and strapped to the bed to prevent them from injuring themselves or others.

Why a snug fit is important

A tight-fitting restraint helps to minimize body movement in a crash. To this end, some seat belts are connected to pretensioning devices that use pyrotechnical charges to tighten the belt around the restrained individual's body when a crash is detected. Many of these restraints are triggered by the same sensor signal that triggers an airbag to deploy.[8]

Not all restraints are created equal

In general, restraints are only effective if they are actually used, fit the person wearing the restraint, and are in good, working condition. To this end, restraints should:

- Be intuitive to engage and disengage.

- Prevent unintended release.

- Be adjustable to ensure a snug fit.

- Be suited for use by people of different sizes.

- Be durable in the intended use environment(s).

Make PPE available and usable

Principle

Personal protective equipment (PPE), worn to minimize exposure to hazards, should be readily accessible and available, easy to don, and comfortable to wear.

What to wear to work today?

The term personal protective equipment (PPE) refers to standalone items that, when worn, reduce one's vulnerability to workplace hazards. PPE includes items such as gloves, goggles, earmuffs, safety shoes, respirators, and entire body suits that people don to shield themselves. The availability of such PPE is as important as its effectiveness. For example, eye protection against flying debris might work quite well, but it doesn't help if the PPE is not on hand when needed or is not worn due to fit and comfort issues.

Make PPE readily available

PPE manufacturers have gone to great lengths to make their products available to those in need, for both for practical and commercial reasons. Dispensers placed near or at the point of use encourage people to don PPE, such as rubber gloves. The idea is to make it convenient so that people will use it.

Some hard hats are designed to be stacked in locations where many people might face an urgent need to protect their heads, such as in earthquake-prone areas in Japan. Other hard hats are designed to interlock with earmuffs or a face shield, addressing multiple protective needs with one convenient solution. Some PPE manufacturers have even developed PPE holders that a user can wear (e.g., a belt clip that holds noise-protection headphones), or that can be integrated into equipment (e.g., a metal bracket that can be attached to a surgical cart to hold a box of rubber gloves) to make PPE available whenever the user needs it.

 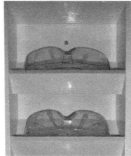 

Hooks and equivalent features enable PPE to be hung on hooks or belts or mounted on a rack.

PPE that is stacked in a convenient place enables people to grab the PPE when needed.

Dispensing systems are a good way to provide single-use items.

Preventing injuries

Many workplace injuries can be prevented by using PPE:

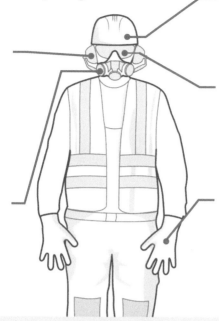

99% of hearing loss, caused by exposure to noises above the recommended limits (see *Principle 4 - Limit sound volume*) can be prevented by wearing proper hearing protection.[1]

15% of adult asthma cases are estimated to be due to work-related exposures that could have been prevented by wearing respiratory protection.[2]

84% of head injuries were sustained by those not wearing proper head protection.[1]

90% of eye injuries in construction workers can be prevented by wearing proper eye protection.[1]

60% of hand and finger injuries can be prevented by protective gloves[3] (and, hand and finger injuries make up **25%** of all workplace accidents).[4]

 Advanced gear used by first responders, such as firefighters, needs to be easily adjustable and donned quickly. The National Fire Protection Association's 1710 standard requires a turnout time (the time between receiving a call and leaving the station) of 80 seconds for fire incidents and 60 seconds for EMS incidents.[5]

Make PPE user-friendly

Opportunities to make PPE more user-friendly (and, therefore, more likely to be used) include the following:

- **Boldly indicate PPE sizes** (e.g., Small, Medium, Large, X-Large) to ensure users can choose the size that fits best and is most comfortable to use.

- **Design PPE to withstand rough handling,** which is likely to occur when used in an emergency scenario, a rush, or a panicked state.

- **Make PPE easy to clean/disinfect** by minimizing seams and crevasses where contamination can collect and be difficult to see and remove (see *Principle 2 - Make things easy to clean*).

- **Make PPE breathable or ventilated** in order to keep wearers comfortable so they are less likely to remove it. For example, protective gloves can be made of a breathable fabric to keep hands cool.

- **Design clasps to provide distinct feedback** when they are opened and closed (e.g., make a sound, produce a "snap" feel, open or close a noticeable gap).

- **Make adjustment mechanisms large enough to manipulate easily,** perhaps while the user is wearing gloves.

- **Make PPE brightly colored and include light-reflecting or -emitting elements** to improve wearers' visibility at nighttime and in smoke-filled environments.

- **Integrate various protective systems into one PPE item,** enabling a single item to provide multiple types of protection.

This protective glove's size is clearly marked (left). This helmet has an integrated face shield and earmuffs, providing head, face, and ear protection in a single piece of equipment (right).

Slow down falling objects

Principle

Falling objects that could strike or crush body parts should incorporate a mechanism that slows down the object's movement. The same principle applies to objects that are rotating, sliding, pivoting, etc.

Newton's contribution to safe design

Some innovations seem so obvious that one wonders why it took so long for them to be developed. "Rollaboard" carry-on suitcases, which were invented in 1987, are one example.[1] Intermittent windshield wipers, which were patented in 1967, are another.[2] So, why did it take until 1995 for the soft close toilet seat to come along?[3] See more on the soft close toilet seat on the following page.

Historically, many products have incorporated components that could drop onto and injure people. Today, in many cases, people are protected by mechanisms that prevent components from dropping too quickly. Slowing down objects (i.e., damping) can reduce impact forces and give people more time to get out of the way of a falling object.

Different types of damping mechanisms

- **Counterweights** are often used as part of pulley systems, such as those used to move theatre stage sets and to give cranes a mechanical advantage. Counterweights balance heavy objects and help users hoist, suspend, and lower such objects slowly and safely without having to fight against heavy gravity loads.

- **Fluid viscous dampers** are filled with small amounts of—you guessed it—highly viscous fluid (e.g., oil). When a heavy object drops, the fluid creates friction and resistance against the cartridge's interior components (e.g., springs, pistons), in turn slowing the heavy object.

- **Hydraulic pistons (struts)** are contained within a cylinder surrounded by fluid or gas. As a heavy object moves or drops, it creates pressure on the piston and forces it to the other side of the cylinder, thereby displacing the surrounding fluid or gas and producing resistance.

Counterweight　　　**No counterweight**

Truck tailgates can be equipped with a damping mechanism to ensure the tailgate drops slowly.

Damping mechanisms are everywhere

Car and van buyers seem to have become enamored with slow, self-closing tailgates. Self-closing tailgates are not only convenient, moving at the touch of a button or the kick of a foot under a sensor; they are also safer because they close slowly. Most self-closing tailgates also stop if they sense motion or significant resistance, like that created by the presence of a child or dog.

Many newer vehicles come equipped with tailgates that self-close at the touch of a button or via a kick sensor; the tailgates close slowly and are designed to stop if they sense motion or resistance.

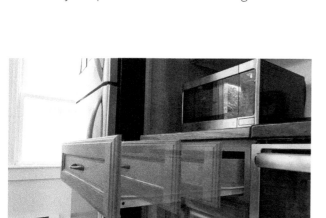

Some kitchen drawers feature soft-close mechanisms that help prevent people from slamming their fingers in the drawer.

Other examples of dampers can be found in the home. Some newer kitchen drawers and cabinets utilize pistons or springs, enabling them to close softly and quietly (sometimes called "anti-slam"), regardless of the initial force applied by a person (within reason). The technology helps prevent people from crushing their fingers in a drawer or cabinet—and from revealing their late-night snacking habits to others in the household.

And finally, society has embraced the "soft close" toilet seat. Undoubtedly, the slow motion of the seat, which can take a leisurely couple of seconds to close, has helped to prevent injuries. What injuries can be caused by a toilet seat? Well, it's fairly obvious why a rapidly closing toilet seat could cause an injury to young males.[3] Also, both boys and girls can get their heads and hands banged up by a dropped toilet seat.

It's all falling into place

In the early 1980s, the US Consumer Products Safety Commission developed a voluntary standard suggesting that toy chest manufacturers use safety hinges that prevent lids from falling forcefully onto children's heads and necks—the reported cause of at least 21 child deaths.[4] Now, toy chests are required to have features that hold the lid in position and require pressure to close it further.[5] These simple features help ensure that children do not get hit on the head or get their fingers crushed by a slamming lid.

Eliminate pinch points

Principle

Get rid of or protect against product elements that might pinch or crush body parts.

What are pinch points?

Pinch points are exactly what they sound like: places where a body part (e.g., hands, feet), or even an entire person, can be compressed by moving parts. Examples include the gap that narrows as a door closes, the intersecting parts of a scissor jack, the crossing structural elements of a stroller, or the place where gears and chains meet.

Pinch points often generate high amounts of pressure due to their "lever nature." As a result, serious damage can occur to a body part caught in a pinch point. Sometimes, the damage is below the skin and can affect tissue, organs, muscles, and bones. Muscle and tissue damage can occur because a pinching accident occludes blood flow. The outcome might be numbness and paralysis of the circulation-deprived area. Additionally, the force of the impact can remove one or more skin layers, leaving an open wound. This can produce an increased chance of infection, which can even lead to amputation in extreme cases.[1]

Drawers create potential pinch hazards that are especially harmful to a child's small hands and fingers. One way to minimize harm is to install "soft close" drawer glides that slowly close when nearing their end position.[2] (Model photo: Colourbox.com)

In 2016, the US Consumer Product Safety Commission issued a recall of more than 29,000 strollers, in part due to the stroller's folding side hinge pinching users' hands with enough force to break the skin.[3] (Photo: Colourbox.com)

How to eliminate pinch points

When possible, pinch hazards should be eliminated through design.

House hazardous components internally

One solution is to cover pinch points. One could debate whether there is any functional difference between a component housing and a protective guard. However, a protective guard is usually placed over an external component, which makes it vulnerable to removal.

The pinching force between the metal chain and a pointy sprocket of a bike chain ring can easily pierce through clothing and skin. A gear case or chain guard (pictured above) encloses the top of the chain and chain ring to protect the cyclist from such harm.

Eliminate pinch hazards

Another way to eliminate a pinch hazard is to opt for a mechanism that achieves the same mechanical function without the need for components to move in a way that could catch a body part. For example, a hydraulic or pneumatic jack could replace a scissor-type lifting mechanism.

A traditional scissor-type car jack can pinch and/or crush a body part if it suddenly collapses. Pictured above, a pneumatic air bag car jack eliminates the pinch point. (Photo: Colourbox.com)

Reduce space between components

Reduce the space between moving parts so that the vulnerable body part cannot fit between them. For example, a gap near moving components could be made so narrow that even a finger cannot fit within the gap. Any residual pinch hazard would likely cause minor injury rather than deep tissue damage.

Flanges extending from the cutting mechanism of a paper trimmer prevent fingers from sliding underneath, thereby minimizing the risk of lacerations. Watch out for a paper cut, though!

Add warning labels

Lastly, if pinch points cannot be eliminated, consider using preventive warning labels that illustrate the type of pinch point or indicate to keep body parts away from moving parts (see *Principle 79 - Add conspicuous warnings*). This is especially relevant for components that can move suddenly and surprise those nearby.

Sometimes, pinch points are inevitable and there is no viable mitigation aside from adding warning labels. Such labels should be easy to interpret, clearly visible, and ideally located near the pinch hazard.

Enable sterilization

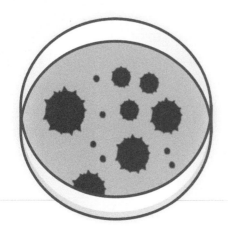

Principle

To prevent infection and transmission of harmful viruses due to a buildup of microbes, make sure products are designed to enable and facilitate sterilization.

Why should products be sterilized?

Microbes are considered the oldest form of life on Earth, and they are practically everywhere. Bacteria and fungi are just two of the many forms that exist today and whose ancestry can be traced back billions of years.[1]

From a human-centric standpoint, some microbes are essential to life, playing a key role in digestion and boosting the immune system.[2] But, other microbes can pose a threat, causing infections such as pneumonia. In 2015 alone, pneumonia claimed 51,811 lives.[3] Coupled with the flu, the Centers for Disease Control and Prevention estimated the duo to be the 8th leading cause of death in the US in 2013.[4] That is why certain products need to be sterilized.

Is this clean?

Where might you find a menagerie of potentially harmful microbes? Try cellphones, kitchen sponges, toilet seats, and men's ties. Men's ties, for example, get loaded with microbes due to sneezes, contact with other objects, and infrequent laundering. That is why some hospitals now prohibit healthcare workers from wearing them in the workplace. Notably, these kinds of contaminated products pose a limited threat to healthy individuals because the microbes are usually the same as those already in the atmosphere and in people's bodies.

Meanwhile, there is a host of products that need to be very clean or even sterile so as to not pose a hazard to users. The principle "Make things easy to clean" addresses ways to make things easy to clean, so the focus here will be on sterilization.

 Viruses are some of the smallest microbes, measuring as little as a few dozen nanometers.[5] For reference, 1 nanometer equals 1 billionth of a meter.

Steam it up

Two common ways to sterilize a physical object are to heat it up, often using steam, and to expose it to ultraviolet (UV) light. Autoclaves sterilize items by exposing them to steam for a certain timeframe at a specific temperature and pressure. Germicidal lamps sterilize items by exposing them to UV-C (short-wavelength UV) light that damages microorganisms' DNA, preventing the microorganisms from replicating. For example, UV lamps installed in pool recirculation pipes use UV-C to kill microbes in swimming pool water. Additional sterilization methods include chemical exposure (using fluid and gases), filtration, and ionizing radiation.[6]

Sterilization can be performed at a small or large scale using tabletop and room-sized machines dedicated to the purpose. In fact, the largest autoclave built held an internal volume of 82,000 ft³, designed to sterilize Boeing 787 Dreamliner passenger aircraft components.[7] In-line electron beam sterilization systems are used in pharmaceutical production, ensuring the safety of products such as pre-filled syringes.[8] Some products undergo a "belt and suspenders" approach, using more than one sterilization method. For example, some machines that produce water for dialysis use heat, filtration, and UV light to ensure the water is free from microbes and bacteria.[9]

Autoclaves use steam to sterilize medical equipment and supplies.

Endoscopes are notorious for causing serious, and even fatal, infections due to improper sterilization. Although part of the endoscope sterilization process is automated, it is largely a manual process, relying on users to perform hundreds of detailed steps correctly to ensure all microbes have been eliminated—a challenge for even the most diligent person.

Things to consider

Here are a few basic considerations for effective sterilization:

- **Consider pretreatment cleaning**
  In some cases, pretreatment cleaning might be needed to ensure effective sterilization. For example, for UV light to be effective, the UV light must reach the microbes. Pretreatment cleaning can remove physical debris that would otherwise block the light and prevent sterilization.

- **Consider automation**
  Determine whether the sterilization process can be automated to reduce the risk that a user fails to sterilize the product properly.

- **Choose the right materials**
  Certain polymers subject to electron beam sterilization can degrade over time due to ionization.[10] Therefore, choose materials based on their sterilization tolerance properties, and ensure that devices can endure the cleaning materials used in the expected use environments.

- **Ensure air can be removed**
  When gases will be used for sterilization purposes, ensure that air within the product can be completely removed and replaced with the microbe-killing gas (e.g., ethylene oxide).[11]

- **Eliminate seams and pockets**
  When fluids will be used to sterilize a product, eliminate seams and pockets that might collect the fluid rather than enabling it to drain away.

Minimize repetitive motion

Principle

Avoid making people perform the same physical task repeatedly, and optimize physical motions to reduce stress.

Why is repetitive motion dangerous?

Repetitive motion injuries are serious medical problems. People who incur them might experience intense pain and be disabled for long periods of time, if not permanently. The injuries are usually the gradually intensifying result of performing the same physical movements over and over again.

Potential repetitive motion injuries

Perhaps the most well-known problem is carpal tunnel syndrome, affecting the wrists where the median nerve passes through a small anatomical structure called the carpal tunnel. In addition to intense pain, carpal tunnel syndrome symptoms include finger tingling and numbness, reduced range of hand and finger motion, and reduced ability to apply force with the hand.[1]

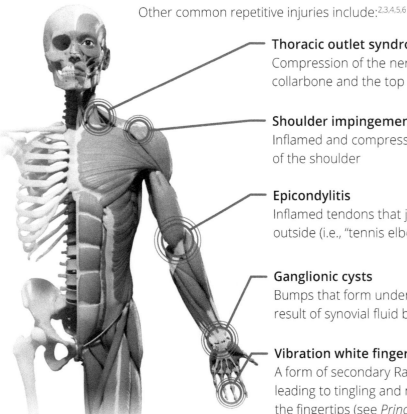

Other common repetitive injuries include:[2,3,4,5,6]

Thoracic outlet syndrome
Compression of the nerves and blood vessels between the collarbone and the top rib

Shoulder impingement syndrome
Inflamed and compressed tendons binding with the bony structure of the shoulder

Epicondylitis
Inflamed tendons that join the forearm muscles on the elbow's outside (i.e., "tennis elbow") or inside (i.e., "golfer's elbow")

Ganglionic cysts
Bumps that form under the skin, typically around the wrist, as a result of synovial fluid buildup

Vibration white finger (VWF)
A form of secondary Raynaud phenomenon: impaired circulation leading to tingling and numbness, ultimately causing whitening of the fingertips (see *Principle 60 - Reduce (or isolate from) vibration*)

Preventing repetitive motion injuries

Reducing the risk of repetitive motion injuries sometimes means changing how people work, such as rotating manual tasks so that workers are not performing the same task for long periods of time. Another approach is to give people breaks between performing the same manual task so that the body has a bit of recovery time—time for blood to circulate and re-oxygenate muscles and other tissues.

But, product designers can also help by implementing ergonomic design guidelines, such as:[7]

- **Facilitating neutral body positions**
  Enable users to maintain as much of a neutral body position as possible when using the product, rather than forcing them to assume strained positions.

- **Limiting application force**
  Reduce the amount of force users must apply to hold or use the product.

- **Minimizing repeated steps**
  Reduce the number of identical steps users must perform to complete an action.

- **Utilizing automation**
  Add mechanisms that minimize the need to apply a constant force for an extended period of time, or automate functions that otherwise would require the user to do the same thing many times.

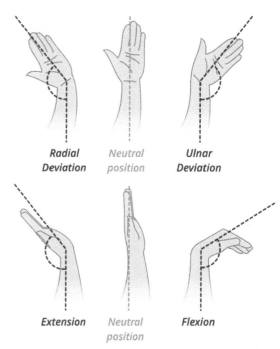

| Radial Deviation | Neutral position | Ulnar Deviation |
| Extension | Neutral position | Flexion |

Knowledge of the body's neutral position should inform design requirements and help designers to minimize strain during repetitive motion.

 For more ergonomic design guidelines, *see Principle 24 - Make it ergonomic.*

Know your users

Demographic and physical characteristics can make a product's users particularly vulnerable to repetitive motion injuries. For example:

- **Age**
  Older individuals are more vulnerable than younger individuals to repetitive motion injuries.[8]

- **Gender**
  Women are three times more likely than men to develop carpal tunnel syndrome.[1]

- **Handedness**
  Left-handed individuals are more likely to develop repetitive motion injuries when using products optimized for right-handed use.[9]

Simple automations advance safety

You might assume that automation requires the use of robotics and artificial intelligence, but sometimes, simple automation mechanisms can provide significant safety benefits. A good example is an electronic pipette, which enables users to fill and empty a pipette with a single button press. This simple advancement reduces the potential for hand and shoulder injuries from repeated pipetting (a frequent activity for many lab technicians) because it greatly reduces the force required to dispense liquid and eject pipette tips.[10]

Exemplar 10
Stretcher

Emergency personnel (e.g., Emergency Medical Technicians) who use stretchers to transport patients need to focus on the patient, rather than on operating the stretcher. As such, it is essential that stretchers are maximally intuitive and inherently safe.

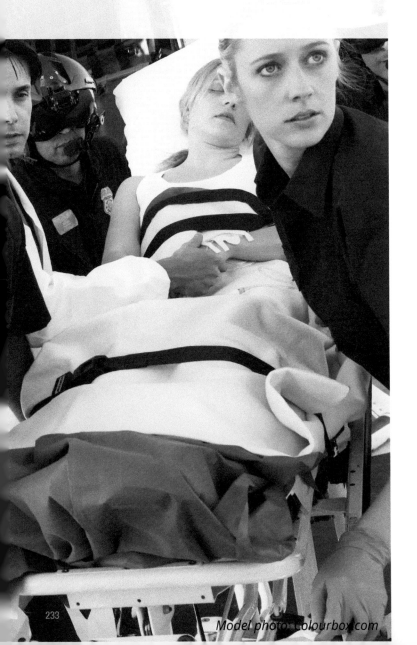

Provide restraints
Principle 95 - pg. 221

Four-point restraint and lap belts secure the patient during transport (which might be bumpy at times). The restraints are adjustable to accommodate patients of various sizes.

Encourage safe lifting / Minimize repetitive motion
Principle 34 - pg. 87 /
Principle 100 - pg. 231 /

Hydraulic system raises and lowers the stretcher with the press of a button, helping users avoid musculoskeletal injuries caused by lifting and lowering heavy loads.

Include brakes
Principle 94 - pg. 219

Brakes keep the stretcher in place while raising and lowering the patient, as well as when the stretcher needs to remain stationary on an inclined surface. The foot-operated brakes enable emergency personnel to keep their arms and hands free for other safety-related purposes.

Model photo: colourbox.com

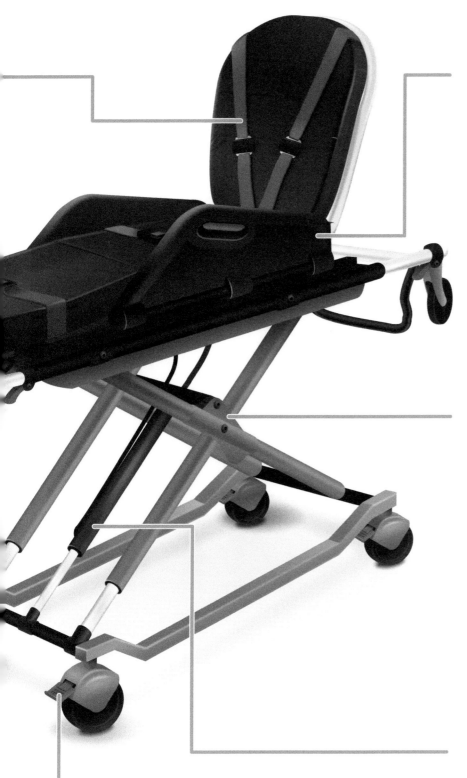

Prevent falls

Principle 39 - pg. 97

Side rails provide emergency personnel with an additional gripping point, as well as reduce the risk of a patient falling out. But, the side rails can be folded down to maximize patient access and care at appropriate times.

Eliminate pinch points

Principle 98 - pg. 227

Streamlined x-frame and lift mechanism eliminate significant gaps between components that could pinch and/or entrap items (e.g., intravenous fluid lines, cables) and body parts.

Slow down falling objects

Principle 97 - pg. 225

Damping mechanism prevents the stretcher from collapsing quickly and suddenly, helping reduce the risk of injury for both the patient and emergency personnel.

Endnotes

Principle 1: Provide stabilization
References:

1. United States, Consumer Product Safety Commission. "Product Instability or Tip-Over Injuries and Fatalities Associated with Televisions, Furniture, and Appliances: 2014 Report." Information on injuries and fatalities caused by televisions, furniture and appliances tip-overs, 2014. https://www.cpsc.gov/s3fs-public/InstabilityorTipoverReport2014Stamped_0.pdf.

Photo credits:

Bicycle – courtesy of Kimmy Ansems

Crane outriggers – courtesy of Michael Wiklund

Principle 2: Make things easy to clean
References:

1. Thompson, Desire. "A McDonald's Employee Was Fired for Exposing Moldy Ice Cream Trays on Twitter." *Vibe*, 28 July 2017, www.vibe.com/2017/07/mcdonalds-employee-fired-moldy-ice-cream-machine-photos/.

2. "Olympus, Fujifilm and Pentax Warned by FDA for Compliance Failures." *MedTech Intelligence*, 12 Mar. 2018, www.medtechintelligence.com/news_article/olympus-fujifilm-pentax-warned-fda-compliance-failures/.

3. Eisler, Peter. "Deadly Bacteria on Medical Scopes Trigger Infections." *USA Today*, 19 Mar. 2015, www.usatoday.com/story/news/2015/01/21/bacteria-deadly-endoscope-contamination/22119329/.

4. United States, Food and Drug Administration. "Reprocessing Medical Devices in Health Care Settings: Validation Methods and Labeling Guidance for Industry and Food and Drug Administration Staff." Reprocessing Medical Devices in Health Care Settings: Validation Methods and Labeling Guidance for Industry and Food and Drug Administration Staff, 2017.

Photo credits:

Bronchoscope – Håkon Olav Leira / CC-BY-SA-3.0

Moldy tray – courtesy of Nick Evans

Man looking into scope – Bartlett, Linda / Public Domain

Principle 3: Eliminate small parts from kids' products
References:

1. United States, Consumer Product Safety Commission, Office of Compliance. "Small Parts Regulations: Toys and Products Intended for Use by Children under 3 Years Old," https://www.cpsc.gov/Global/Business-and-Manufacturing/Business-Education/Business-Guidance/Small%20Parts/Regsumsmallparts.pdf, Jan. 2001.

2. United States Public Interest Research Group, Education Fund, et al. "Trouble in Toyland: The 29th Annual Survey of Toy Safety," uspirg.org/sites/pirg/files/reports/USPIRG_Trouble_in_Toyland_2014.pdf, Nov. 2014.

3. UL, "Guide to International Toy Safety Requirements." https://industries.ul.com/wp-content/uploads/sites/2/2014/01/InternationalToyPamphlet_12132013_TBLOD_Web.pdf, Dec. 2013.

4. CPSC Blogger. "Small Parts: What Parents Need to Know." *On Safety*, Consumer Product Safety Commission, 20 Dec. 2011, onsafety.cpsc.gov/blog/2011/12/20/small-parts-what-parents-need-to-know/.

5. "A Dangerously Tasty Treat: The Hot Dog Is a Choking Hazard," *Raising Healthy Children*, www.hopkinsmedicine.org/health/articles-and-answers/healthy-children/a-dangerously-tasty-treat-the-hot-dog-is-a-choking-hazard.

6. Cyr, Claude. "Preventing Choking and Suffocation in Children." *Paediatrics & Child Health*, vol. 17, no. 2, 2012, pp. 91–92, doi:10.1093/pch/17.2.91.

7. Lumsden, Amy J., and Jamie G. Cooper. "The Choking Hazard of Grapes: A Plea for Awareness." *Archives of Disease in Childhood*, vol. 102, no. 5, 2016, pp. 473–474., doi:10.1136/archdischild-2016-311750.

Principle 4: Limit sound volume
References:

1. "Loud Noise Dangers." *American Speech-Language-Hearing Association*, ASHA, www.asha.org/public/hearing/Noise/.

2. "Solar Retinopathy." *Eyecare Trust*, www.eyecaretrust.org.uk/view.php?item_id=104.

3. "Noise-Induced Hearing Loss." *National Institute of Deafness and Other Communication Disorders*, U.S. Department of Health and Human Services, 7 Feb. 2017, www.nidcd.nih.gov/health/noise-induced-hearing-loss.

4. "Laboratory Safety Noise." *OSHA Fact Sheet*, OSHA, 2011, www.osha.gov/Publications/laboratory/OSHAfactsheet-laboratory-safety-noise.pdf.

5. United States, Congress, Occupational Safety and Health Administration. "§1926.52 Occupational Noise Exposure." *§1926.52 Occupational Noise Exposure*, e-CFR, 5 Apr. 2018. www.ecfr.gov/cgi-bin/text-idx?SID=f9e6850a9fc1328e19d5f11c8baaa029&mc=true&node=se29.8.1926_152&rgn=div8.

6. Dangerous Decibels. "Noise Induced Hearing Loss (NIHL)." *Dangerous Decibels*, dangerousdecibels.org/education/information-center/noise-induced-hearing-loss/.

7. "5.2.1 Noise Reduction and Sound Absorption." *Environmental Pollution Control*, by Jingling Liu et al., Walter De Gruyter GmbH, 2017, pp. 142–144.

8. "The Science behind the Noise Cancelling Technology." *TechTalk*, TechTalk, 11 June 2013, techtalk.currys.co.uk/audio/headphones/the-science-behind-the-noise-cancelling-technology/.

9. The Editors of Encyclopaedia Britannica. "Muffler." *Encyclopædia Britannica*, Encyclopædia Britannica, Inc., 30 Sept. 2008, www.britannica.com/technology/muffler-engine-part.

10. Doughty, Steve. "Nearly Silent Electric or Hybrid Cars 'Are a Risk to Pedestrians': Walkers 40% More Likely to Be Involved in Accident ." *Daily Mail Online*, Associated Newspapers, 25 Mar. 2015, www.dailymail.co.uk/sciencetech/article-3011957/Nearly-silent-electric-hybrid-cars-risk-pedestrians-Walkers-40-likely-involved-accident.html.

Principle 5: Include pads
References:

1. "Helmets: How They Work, and What Standards Do." *Helmets.org*, 8 Oct. 2017, www.helmets.org/general.htm.

2. Motorcycle USA Staff. "Motorcycle Helmets Product Guide." Motorcycle USA, *MotoUSA.com*, 19 Nov. 2007, www.motorcycle-usa.com/2007/11/article/motorcycle-helmets-product-guide/.

3. "Evolution of Car Safety." *Allianz Australia*, 2018 Allianz Australia Limited, www.allianz.com.au/car-insurance/news/evolution-of-car-safety. ABN 21 000 006 226.

Principle 6: Make it buoyant
References::

1. "Dow Floating Billets." *Dock Builders Supply*, www.dockbuilders.com/dow-styrofoam-billets.htm.

2. Philpott, Sarah. "A Warning for All Parents: Don't Use Water Wings in the Swimming Pool." *Her View from Home*, 26 June 2016, herviewfromhome.com/a-warning-for-all-parents-dont-use-water-wings-in-the-swimming-pool/.

3. Dzierzak, Lou. "How to Survive an Avalanche: Skier's Air Bag." *Scientific American*, 29 Jan. 2013, www.scientificamerican.com/article/survive-an-avalanche-skier-air-bag/.

4. "Life Jacket Learning: PFD Selection Guide." *Jamestown Distributors*, https://www.jamestowndistributors.com/userportal/document.do?docId=1032.

5. "Life Jacket Testing: It's a Safety Thing." *UL*, 2018, www.ul.com/inside-ul/life-jacket-verification-its-a-safety-thing/.

Principle 7: Temper the glass

References:

1. Kumar, Lalit, and Tom Norton. "Difference between Gorilla Glass and Tempered Glass." *TechWelkin*, 31 July 2017, techwelkin.com/gorilla-glass-vs-tempered-glass.

2. "How It's Made." *Corning | Gorilla® Glass*, Corning Incorporated, www.corning.com/gorillaglass/worldwide/en/technology/how-it-s-made.html.

3. "ANSI Z97.1." *IHS Markit Standards Store | Engineering & Technical Information, Acoustical Society of America,* 2018, global.ihs.com/doc_detail.cfm?&rid=ASA&item_s_key=00010381&item_key_date=841231&input_doc_number=&input_doc_title=.

Principle 8: Provide tactile feedback

References:

1. Jensen, Christopher. "Anton Yelchin's Death Highlights a Known Issue with Jeeps." *The New York Times,* 21 June 2016, www.nytimes.com/2016/06/22/business/anton-yelchins-death-highlights-a-known-issue-with-jeeps.html.

2. Abrahamson, Eric, and David H. Freedman. *A Perfect Mess: The Hidden Benefits of Disorder: How Crammed Closets, Cluttered Offices and on-the-Fly Planning Make the World a Better Place.* Weidenfeld & Nicolson, 2006.

3. "What Is That Yellow Line on the Sidewalk?" *Digital Journal*, www.digitaljournal.com/blog/21332.

Principle 9: Make it fire resistant

References:

1. "Fire 101: The Dangers of Fire?" *Purdue University - Fire Department,* 2014, www.purdue.edu/ehps/fire/fire-101.html.

2. "Children's Sleepwear Regulations." *CPSC.gov*, www.cpsc.gov/Business--Manufacturing/Business-Education/Business-Guidance/Childrens-Sleepwear-Regulations.

3. Haberman, Clyde. "A Flame Retardant That Came with Its Own Threat to Health." *The New York Times*, 3 May 2015, www.nytimes.com/2015/05/04/us/a-flame-retardant-that-came-with-its-own-threat-to-health.html.

4. "Nomex." *6 Minutes for Safety*, wildfirelessons.connectedcommunity.org/HigherLogic/System/DownloadDocumentFile.ashx?DocumentFileKey=b4ebac61-c423-465d-8235-0d3055396e70&forceDialog=0.

5. Coxworth, Ben. "Fire Blanket uses spaceship tech to protect forest-Firefighters." New Atlas - New Technology & Science News, *New Atlas,* 27 Oct. 2014, newatlas.com/sunseeker-fire-blanket/34442/.

6. Swider, Matt. "Here's Why the Samsung Galaxy Note 7 Batteries Caught Fire and Exploded." *TechRadar*, 23 Jan. 2017, www.techradar.com/news/samsung-galaxy-note-7-battery-fires-heres-why-they-exploded.

7. Firth, Michael. "Stanford University Researchers Have Produced a "Self-Extinguishing" Lithium-Ion Battery." *Neowin*, 16 Jan. 2017, www.neowin.net/news/stanford-university-researchers-have-produced-a-self-extinguishing-lithium-ion-battery.

8. "New Science | Fire Safety." *UL New Science*, 2016, newscience.ul.com/firesafety.

Principle 10: Use non-toxic materials

References:

1. Hernandez, Miguel. "Your Health, the Environment and Wooden Telephone Poles." *Patch Media*, 13 Dec. 2012, patch.com/new-york/ossining/bp--your-health-the-environment-and-wooden-telephone-poles.

2. King, Daniel. "Molded Asbestos Plastic Products: Brands & Products." Edited by Walter Pacheco, *Asbestos.com*, 15 Jan. 2018, www.asbestos.com/products/general/plastics.php.

3. Petre, Alina. "What Is BPA and Why Is It Bad for You?" *Healthline*, Healthline Media, 23 Mar. 2016, www.healthline.com/nutrition/what-is-bpa.

4. Bilbrey, Jenna. "BPA-Free Plastic Containers May Be Just as Hazardous." *Scientific American*, 11 Aug. 2014, www.scientificamerican.com/article/bpa-free-plastic-containers-may-be-just-as-hazardous/.

5. "The Frank R. Lautenberg Chemical Safety for the 21st Century Act." *EPA.gov*, 17 May 2018, www.epa.gov/assessing-and-managing-chemicals-under-tsca/frank-r-lautenberg-chemical-safety-21st-century-act.

6. "ASTM E595 - 15 Standard Test Method for Total Mass Loss and Collected Volatile Condensable Materials from Outgassing in a Vacuum Environment." *ASTM International - Standards Worldwide*, www.astm.org/Standards/E595.htm.

7. "Toxicity of Plastics." *Blastic*, www.blastic.eu/knowledge-bank/impacts/toxicity-plastics/.

8. Steinemann, Anne. "Ten Questions Concerning Air Fresheners and Indoor Built Environments." *Building and Environment*, vol. 111, pp. 279–284., doi.org/10.1016/j.buildenv.2016.11.009.

9. Jaslow, Ryan. "New Car Smell Is Toxic, Study Says: Which Cars Are Worst?" *CBS News*, 15 Feb. 2012, www.cbsnews.com/news/new-car-smell-is-toxic-study-says-which-cars-are-worst/.

10. Travers, Jim. "Autos: Is New-Car Smell Bad for Your Health?" *BBC*, 15 Mar. 2016, www.bbc.com/autos/story/20160315-is-new-car-smell-bad-for-your-health.

Exemplar 1: Car seat

Principle 11: Lock touchscreens

References:

1. United States Patent: 8046721, USPTO, 25 Oct. 2011, patft.uspto.gov/netacgi/nph-Parser?Sect1=PTO1&Sect2=HITOFF&d=PALL&p=1&u=%2Fnetahtml%2FPTO%2Fsrchnum.htm&r=1&f=G&l=50&s1=8%2C046%2C721.PN.&OS=PN%2F8%2C046%2C721&RS=PN%2F8%2C046%2C721.

2. Kahney, Leander. "The Inside Story of the IPhone's 'Slide to Unlock' Gesture." *Cult of Mac*, 30 June 2017, www.cultofmac.com/490394/iphone-slide-to-unlock-bas-ording/.

3. Roemmele, Brian. "How Does Apple's New Face ID Technology Work?" *Forbes*, 13 Sept. 2017, www.forbes.com/sites/quora/2017/09/13/how-does-apples-new-face-id-technology-work/#e4a16de2b7f4.

4. Bomey, Nathan. "Which Cars Are Most Distracting? AAA Study Reveals Offenders." *USA Today*, Gannett Satellite

Information Network, 5 Oct. 2017, www.usatoday.com/story/money/cars/2017/10/05/aaa-distracted-driving-infotainmentstudy/734677001/.

Photo credits:

Cardiac arrest scene – from Pinsdaddy / Photo credit: Azureedge.net

Principle 12: Make buttons large

References:

1. Association for the Advancement of Medical Instrumentation. *ANSI/AAMI HE75, 2009 Edition: Human Factors Engineering—Design of Medical Devices*. American National Standard, 2009.

2. U.S. Nuclear Regulatory Commission, Office of Nuclear Regulatory Research. *Human-System Interface Design Review Guidelines*, NUREG-0700, Rev. 2, 2002.

3. "Fitts's Law: The Importance of Size and Distance in UI Design." *The Interaction Design Foundation*, 2016, www.interaction-design.org/literature/article/fitts-s-law-the-importance-of-size-and-distance-in-ui-design.

Principle 13: Use consistent units of measure

References:

1. Hotz, Robert Lee. "Mars Probe Lost Due to Simple Math Error." *Los Angeles Times*, 1 Oct. 1999, articles.latimes.com/1999/oct/01/news/mn-17288.

2. Williams, David R. "Mars Climate Oribiter." *NASA Space Science Data Coordinated Archive*, 21 Mar. 2017, nssdc.gsfc.nasa.gov/nmc/spacecraftDisplay.do?id=1998-073A.

3. "Accident Description." *Aviation Safety Network*, 23 July 1983, aviation-safety.net/database/record.php?id=19830723-0.

4. Witkin, Richard. "Jet's Fuel Ran out after Metric Conversion Errors." *The New York Times*, 30 July 1983, www.nytimes.com/1983/07/30/us/jet-s-fuel-ran-out-after-metric-conversion-errors.html.

5. "Your Child's Size and Growth Timeline." *BabyCenter*, Oct. 2016, www.babycenter.com/0_your-childs-size-and-growth-timeline_10357633.bc.

6. Bokser, Seth J. "A Weighty Mistake." *PSNet: Patient Safety Network*, Mar. 2013, psnet.ahrq.gov/webmm/case/293/a-weighty-mistake.

7. "Medication Errors: Significance of Accurate Patient Weights." *Pennsylvania Patient Safety Advisory*, vol. 6, no. 1, Mar. 2009, pp. 10–15.

Photo credits:

Boeing 767_233 C-GAUN – used with permission from Ken Spicer

Principle 14: Ensure strong label-control associations

References:

1. Caruso, Catherine. "The Problems with Poor Ballot Design." *Scientific American*, 7 Nov. 2016, www.scientificamerican.com/article/q-a-with-philip-kortum/.

2. Tidwell, Jenifer. "The Palm Beach Ballot Fiasco.", 8 Nov. 2000, www.mit.edu/~jtidwell/ballot_design.html.

3. "A Dominant Hemisphere for Handedness and Language?" *ScienceDaily*, CNRS, 4 July 2014, www.sciencedaily.com/releases/2014/07/140704134633.htm.

Photo credits:

Ballot – from Wikimedia Commons / Public Domain

HAMILTON-C1 ventilator – used with permission from Hamilton© www.hamilton-medical.com

Principle 15: Provide reminders

References:

1. "5 Cool Pieces of Skydiving Gear You Didn't Know About." *Skydive Long Island*, 16 Jan. 2007, www.skydivelongisland.com/about/articles/5-cool-pieces-of-skydiving-gear-you-didn-t-know-about/.

2. "What's All The Flap About?" *Plane & Pilot Magazine*, www.planeandpilotmag.com/article/whats-all-the-flap-about/#.Wrf43BPwbfY.

3. Pugel, Anne E., et al. "Use of the Surgical Safety Checklist to Improve Communication and Reduce Complications" *J Inject Public Health*, 2015, https://www.ncbi.nlm.nih.gov/pmc/articles/PMC4417373/.

4. Shifflett, Tammy Shifflett. "Continuous Glucose Monitoring: Everything You Need to Know." *TheDiabetesCouncil.com*, 25 Mar. 2018, www.thediabetescouncil.com/continuous-glucose-monitoring-everything-you-need-to-know/.

5. Luthra, Shefali. "Doctors Are Overloaded with Electronic Alerts, and That's Bad for Patients." *The Washington Post*, WP Company, 13 June 2016, www.washingtonpost.com/national/health-science/doctors-are-overloaded-with-electronic-alerts-and-thats-bad-for-patients/2016/06/10/0cae6b4a-20fa-11e6-9e7f-57890b612299_story.html?utm_term=.57180ff1d605.

Photo credits:

Doctor at computer – from Colourbox.com

Principle 16: Make decimal values distinct

References:

1. Sullivan, Bob. "5 Things to Know about Capnography and Respiratory Distress." *EMS1*, 13 Oct. 2015, www.ems1.com/ems-products/medical-equipment/airway-management/articles/14933048-5-things-to-know-about-capnography-and-respiratory-distress/.

Principle 17: Display real-time data or use time stamps

References:

1. "ANSI/AAMI HE75:2009 Human Factors Engineering: Design of Medical Devices." American National Standard, 21 Oct. 2009.

2. "Automated Monitoring of Noninvasive Blood Pressure (NIBP)." *Medtronic Physio-Control*, Medtronic, Inc., https://www.physio-control.com/uploadedFiles/learning/clinical-topics/3012791-000%20Automated%20Monitoring%20of%20Noninvasive%20Blood%20Pressure%20(NIBP).pdf.

3. Vital Signs Monitor 300 Series: Directions for Use Software Version 1.2X. *Welch Allyn*, 2012, www.welchallyn.com/content/dam/welchallyn/documents/sap-documents/LIT/810-2/810-2222-05LITPDF.pdf.

Photo credits:

Blood glucose meter, car dashboard – courtesy of Kimmy Ansems

Principle 18: Predict hazardous situations

References:

1. "3 Ways to Get Started." *MiniMed 670G Insulin Pump System | World's First Hybrid Closed Loop System*, Medtronic MiniMed, Inc., 2018, www.medtronicdiabetes.com/products/minimed-530g-diabetes-system-with-enlite.

2. Schilling, David Russell. "FireCast Predictive Analytics: Where Is the Next Fire Likely to Be?" *Industry Tap*, 20 Dec. 2014, www.industrytap.com/firecast-predictive-analytics-next-fire-likely/23703.

3. Davies, Alex. "The Very Human Problem Blocking the Path to Self-Driving Cars." *Wired*, Conde Nast, 1 Feb. 2018, www.wired.com/2017/01/human-problem-blocking-path-self-driving-cars/.

Principle 19: Make software secure

References:

1. U.S. Department of Commerce, National Institute of Standards and Technology. *NIST Special Publication 800-63-3, Digital Identity Guidelines, 2017*, https://pages.nist.gov/800-63-3/sp800-63-3.html.

2. "Kovacs, Eduard. "FDA Issues Alert over Vulnerable Hospira Drug Pumps." *SecurityWeek: Information Security News*, Insights & Analysis , 3 Aug. 2015, www.securityweek.com/fda-issues-alert-over-vulnerable-hospira-drug-pumps.

3. Guccione, Darren. "What the Most Common Passwords of 2016 List Reveals [Research Study]." *Keeper Blog*, 13 Jan. 2017, keepersecurity.com/blog/2017/01/13/most-common-passwords-of-2016-research-study/.

4. Kobie, Nicole. "Security vs Usability: It Doesn't Have to Be a Trade-Off." *The Telegraph*, Telegraph Media Group, 28 July 2016, www.telegraph.co.uk/connect/better-business/security-versus-usability-ux-debate/.

Principle 20: Do not require mental calculation

References:

1. Chinn, Steve, and Richard Edmund Ashcroft. *Mathematics for Dyslexics: Including Dyscalculia*. 3rd ed., John Wiley & Sons, Ltd, 2007.

2. Molko, Nicolas, et al. "Functional and Structural Alterations of the Intraparietal Sulcus in a Developmental Dyscalculia of Genetic Origin." *Neuron*, vol. 40, no. 4, Nov. 2003, pp. 847–858., doi:https://doi.org/10.1016/S0896-6273(03)00670-6.

Exemplar 2: Diabetes management software

Photo credits:

Blood glucose test – from Biswarup Ganguly / Wikimedia Commons / CC-BY-3.0

Principle 21: Incorporate automatic feeding mechanism

References:

1. Mehta, Benita. "Machine Guard Violations Can Cause Serious Injury." *ISHN Industrial Safety & Hygiene News*, BNP Media, 1 Oct. 2015, www.ishn.com/articles/102434-machine-guard-violations-can-cause-serious-injury.

2. Fachot, Morand. "Cutting Risks in Human-Robot Interaction | IEC e-Tech | Issue' 02/2012." *IEC e-Tech*, Feb. 2012, https://iecetech.org/index.php/issue/2012-02/Cutting-risks-in-human-robot-interaction.

3. "Safeguarding Equipment and Protecting Employees from Amputations." *Occupational Safety and Health Administration*, U.S. Dept. of Labor, 2007. OSHA 3170-02R 2007.

Photo credits:

Large wood chipper, meat feeding mechanism, packing line button panel – from Colourbox.com

Laser guard – used with permission from Press Brake Safety

Principle 22: Provide illumination

References:

1. "6.15 Lighting." *U.S. General Services Administration*, U.S. Government, 13 Aug. 2017, www.gsa.gov/node/82715.

2. Illuminance - Recommended Light Level. *The Engineering ToolBox*, 2004, www.engineeringtoolbox.com/light-level-rooms-d_708.html.

3. International Electrotechnical Commission. *Medical Electrical Equipment Part 2-41: Particular Requirements for Basic Safety and Essential Performance of Surgical Luminaires and Luminaires for Diagnosis.* NEN-EN-IEC 60601-2-41:2010. Geneva: IEC, 2009.

4. Thompson, Robert Bruce., and Barbara Fritchman Thompson. *Astronomy Hacks: Tips & Tools for Observing the Night Sky.* O'Reilly, 2005.

Photo credits:

Drill – courtesy of Ruben Post

Head lamp – from Senior Airman Christopher Callaway, U.S. Air Force / Public Domain

Lit stairs, traffic lights – from Pixabay / Public Domain

Lit buttons – from user: Torley / Flickr / CC-BY-SA-2.0

Safety vest – from Magnus Mertens / Wikimedia Commons / CC-BY-SA-2.0

Submarine interior – from D. Myles Cullen, Joint Chiefs of Staff / Public Domain

Surgical light – from Swafford, U.S. Air Force / Public Domain

Principle 23: Add a "dead man's switch"

References:

1. Higgins, Chris. "The Dead Man's Switch." *Mental Floss*, Mental Floss, Inc., 30 May 2008, mentalfloss.com/article/18749/dead-mans-switch.

2. Semsel, Craig R. *"Built to Move Millions: Streetcar Building in Ohio."* Indiana University Press, 2008.

3. "Consumer Reports." *The Washington Post*, 19 June 1990, safetyengineeringresources.com/2013FC54. https://www.washingtonpost.com/archive/lifestyle/1990/06/19/consumer-reports/9d8bcf25-2df9-4efd-8f5b-8d3cb4e29c85/?noredirect=on&utm_term=.461c740495db.

4. Sagan, Scott D. *The Limits of Safety: Organizations, Accidents, and Nuclear Weapons*. Princeton University Press, 1993.

5. Shenfield, Stephen D. "Nuclear Command and Control: From Fail-Safe to Fail-Deadly." 5 May 2012, stephenshenfield.net/themes/war-and-disarmament/nuclear-weapons/114-nuclear-command-and-control-from-fail-safe-to-fail-deadly.

Principle 24: Make it ergonomic

<u>References:</u>

1. Profis, Sharon. "5 Ways to Make Your Office Desk More Ergonomic." *CNET*, 1 Jan. 2016, www.cnet.com/how-to/how-to-set-up-an-ergonomic-workstation/.

2. Hedge, Alan. "Neutral Posture Typing." *CUergo Cornell University Ergonomics Web*, 13 June 2015, ergo.human.cornell.edu/AHTutorials/typingposture.html.

3. National Health and Nutrition Examination Survey III: Body Measurements (Anthropometry). 1988, Westat, Inc., Baltimore, MD.

4. *National Health and Nutrition Examination Survey (NHANES) Anthropometry Procedures Manual.* CDC, Jan. 2007, www.cdc.gov/nchs/data/nhanes/nhanes_07_08/manual_an.pdf.

5. Association for the Advancement of Medical Instrumentation. *ANSI/AAMI HE75, 2009 Edition: Human Factors Engineering—Design of Medical Devices,* American National Standard, 2009.

6. Openshaw, Scott, and Erin Taylor. *Ergonomics and Design: A Reference Guide*. Diane Publishing, 2006.

Principle 25: Armor it

<u>References:</u>

1. Rovell, Darren. "Will Kevlar Revolutionize Sports." *CNBC*, 13 Nov. 2009, www.cnbc.com/id/33909937.

2. "Protective K-9 Ballistic Vests (Bullet/Stab)." *Project Paws Alive*, 31 Jan. 2014, projectpawsalive.org/protective-k-9-ballistic-vests/.

3. "Kevlar." *Illustrated Glossary of Organic Chemistry*, Institute for Reduction of Cognitive Entropy in Organic Chemistry, www.chem.ucla.edu/~harding/IGOC/K/kevlar.html.

4. Harris, Tom. "How M1 Tanks Work." *HowStuffWorks Science,* HowStuffWorks, 8 Mar. 2018, science.howstuffworks.com/m1-tank4.htm

5. "How Rugged Are Your Products?" *Eurofins*, MET Laboratories, Inc., 24 May 2018, www.metlabs.com/military/mil-std-810-commercial-product-ruggedization/.

6. "Toughpad." *Panasonic*, 2012, in.panasonictoughbook.asia/computer-product/introducing-the-full-toughbook-range/toughpad.

7. "Types of Radiation: Gamma, Alpha, Neutron, Beta & X-Ray Radiation Basics." *Mirion*, www.mirion.com/introduction-to-radiation-safety/types-of-ionizing-radiation/.

Sergeant using armored laptop – from Staff Sgt. Kenny Holston, U.S. Air Force / Public Domain

Tank – from user: ZStoler / Wikimedia Commons / Public Domain

Principle 26: Incorporate a lockout mechanism

References:

1. Richards, Laura. "9 Things to Know Before You Self-Clean Your Oven." *Reader's Digest*, 17 Aug. 2017, www.rd.com/home/cleaning-organizing/self-clean-oven/.

2. Erbe Elektromedizin GmbH. *VIO® NESSY® System and NESSY Ω® - Safe Use of Return Electrodes*, 2017, www.erbe-med.com/erbe/media/Marketingmaterialien/85800-107_ERBE_EN_NESSY_brochure_Application_of_Patient_Plates_D024810.pdf.

Photo credits:

Car breathalyzer – from Smart Start Inc.

Roller coaster – from Pixabay

Washing machine – courtesy of Rachel Aronchick

Principle 27: Eliminate or limit toxic fumes

References:

1. Lander, Steve. "The Hazards of Photocopier Toner." *Chron*, http://smallbusiness.chron.com/hazards-photocopier-toner-67532.html.

2. Farquhar, Caroline. "Dangers of Ozone in the Office." *Naturallysavvy*, 16 December 2010, naturallysavvy.com/care/dangers-of-ozone-in-the-office.

3. Engleson, Joe. "Problems with Ozone Generators and Ionizers that Produce Ozone." *Air Technology Solutions*, www.airmation.ca/article-ozonegenerators.php.

4. The United States Environmental Protection Agency. "Indoor Air Quality. Volatile Organic Compounds' Impact on Indoor Air Quality." *EPA*, https://www.epa.gov/indoor-air-quality-iaq/volatile-organic-compounds-impact-indoor-air-quality.

5. "VOC Regulations and What They Mean for Manufacturers." *US Coatings*, www.uscoatings.com/blog/voc-regulations-and-what-they-mean-for-manufacturers/.

6. "Natural Gas Smell." *Union Gas*, www.uniongas.com/about-us/safety/gas-smell-safety.

7. Waste Anesthetic Gases: Occupational Hazards in Hospitals. Department of Health and Human Services, Centers for Disease Control and Prevention National Institute for Occupational Safety and Health, www.cdc.gov/niosh/docs/2007-151/pdfs/2007-151.pdf.

8. United States Department of Labor. "Anesthetic Gases: Guidelines for Workplace Exposures." *US Dept. of Labor*, https://www.osha.gov/dts/osta/anestheticgases/.

Photo credits:

Anesthetic Gas Scavenging System – used with permission from BeaconMedæs

Principle 28: Add a horn, whistle, beeper, or siren

References:

1. Privopoulos, Elefterios P., Carl Q. Howard, and Aaron J. Maddern. "Acoustic Characteristics For Effective Ambulance Sirens." *Acoustics Australia* 39.2 (2011): 43.

2. "Horns, Backup Alarms, and Automatic Warning Devices." *Mine Safety and Health Administration - MSHA* , United States Department of Labor, arlweb.msha.gov/STATS/Top20Viols/tips/14132.htm.

3. United States Departmen of Labor - Safety and Health Regulations for Construction. Motor Vehicles, Mechanized Equipment, and Marine Operations. OSHA. 29CFR 1926.601(b)(4)(i-ii). https://www.osha.gov/pls/oshaweb/owadisp.show_document?p_table=STANDARDS&p_id=10768

4. Doughton, Malcom, et al. Stage 2 Design. Cengage Learning EMEA, 2001.

5. ISO 7731:2003 - Ergonomics -- Danger signals for public and work areas -- Auditory Danger Signals, ISO, 2003.

6. "Noise Induced Hearing Loss." *American Hearing Research Foundation*, Oct. 2012, american-hearing.org/disorders/noise-induced-hearing-loss/.

7. Krug, Etienne G. *Hearing Loss Due to Recreational Exposure to Loud Sounds: A Review*. World Health Organization, 2015.

Principle 29: Let users set the pace

References:

1. Parkes, Katharine R., et al. "Work Preferences as Moderators of the Effects of Paced and Unpaced Work on Mood and Cognitive Performance: A Laboratory Simulation of Mechanized Letter Sorting." *Human Factors: The Journal of the Human Factors and Ergonomics Society*, vol. 32, no. 2, Apr. 1990, pp. 197–216., doi:10.1177/001872089003200207.

2. Cepeda, Nicholas J., et al. "Speed Isn't Everything: Complex Processing Speed Measures Mask Individual Differences and Developmental Changes in Executive Control." *Developmental Science*, vol. 16, no. 2, 2013, pp. 269–286., doi:10.1111/desc.12024.

3. Smith, Sandy. "Tesla: Is Safety Sacrificed to Production?" *EHS Today*, 12 June 2017, www.ehstoday.com/safety-leadership/tesla-safety-sacrificed-production.

4. Kohlstedt, Kurt. "Paternoster Lifts: Cyclic Chain Elevators with No Buttons, Doors or Stops." *99% Invisible*, 16 May 2016, 99percentinvisible.org/article/paternoster-lifts-cyclic-chain-elevators-no-buttons-doors-stops/.

5. "Smart Drive Speed Control." *Honda Power Equipment*, powerequipment.honda.com/lawn-mowers/smart-drive-speed-control.

6. Soegaard, Mads. "Hick's Law: Making the Choice Easier for Users." *The Interaction Design Foundation*, Jan. 2018, www.interaction-design.org/literature/article/hick-s-law-making-the-choice-easier-for-users.

Principle 30: Protect against roll-over and tip-over

References:

1. Murphy, Dennis. "Rollover Protection for Farm Tractor Operators." *Penn State Extension*, Aug. 2014, extension.psu.edu/rollover-protection-for-farm-tractor-operators.

2. Edmonston, Phil. *Lemon-Aid New Cars and Trucks 2011*. Dundurn Press, 2010.

3. Hollembeak, Barry. *Classroom Manual for Advanced Automotive Electronic Systems*. Cengage Learning, Inc., 2010.

4. International Electrotechnical Commission. "IEC 60601-1ISO 60601-1: 9.4.2.3 Instability from horizontal and vertical forces." IEC 60601-1ISO 60601-1: Medical electrical equipment – Part 1: General requirements for basic safety and essential performance, Edition 3.1. Aug 2012.

5. Beckmann, Brittany. "IEC 60601-1 Instability Test Break Down for Custom Medical Carts." *HUI*, 13 Oct. 2016, www.medicalcarts.org/blog/iec-60601-1-instability-test-break-down-for-custom-medical-carts.

6. U.S. Consumer Product Safety Commission. "Staff Briefing Package on Furniture Tipover," Sept. 2016. https://www.cpsc.gov/s3fs-public/Staff%20Briefing%20Package%20on%20Furniture%20Tipover%20-%20September%2030%202016.pdf.

7. Kids In Danger and Shane's Foundation. "Furniture Stability: A Review of Data and Testing Results." *Kids in Danger*. 9 Aug 2016. http://www.kidsindanger.org/docs/research/Furniture_Stability_Report_Final.pdf.

8. United States Consumer Product Safety Commission. "1 Child Dies Every Two Weeks: Tipover Dangers." https://www.cpsc.gov/safety-education/safety-guides/furniture-furnishings-and-decorations/1-child-dies-every-two-weeks.

9. Berg, Tom. "Rollover Control: Electronic Stability Technology." *HDT Truckinginfo*, June 2009, www.truckinginfo.com/channel/fleet-management/article/story/2009/06/rollover-control-electronic-stability-technology.aspx.

10. "Car Rollover 101." *Consumer Reports*, Apr. 2014, www.consumerreports.org/cro/2012/02/rollover-101/index.htm.

11. "Rollover Ratings Include Dynamic Test." *IIHS HLDI*, 6 Mar. 2004, www.iihs.org/iihs/sr/statusreport/article/39/3/3.

12. Ballaban, Michael. "Here's Another Horrific Reminder of Why the Moose Test Is a Thing." Jalopnik, Jalopnik.com, 28 June 2016, jalopnik.com/heres-another-horrific-reminder-of-why-the-moose-test-i-1782739580.

Exemplar 3: Tractor
Photo credits:

Tractor – from Pixabay / Public Domain

Principle 31: Make parts move, deform, or disconnect
References:

1. Keilman, John. "This Gadget Really Was a Slam-Dunk." *Chicago Tribune*, 4 Apr. 2005, articles.chicagotribune.com/2005-04-04/news/0504040109_1_dunk-rim-grain-elevator).

2. How to Replace a Snowblower Shear Pin. *Sears PartsDirect,* www.searspartsdirect.com/repair-guide/snowblower/how-to-replace-a-snowblower-shear-pin.html.

3. Asiminei, Aida Georgeta, et al. "A Comparison Study of Different Cyclist Helmet Designs by Finite Element Analysis." *Proceedings of The 2008 Finite Element Workshop,* 2008, Ulm, Germany. Vol. 1. KFO-Wissenschaft, Universitatsklinikum Ulm, ZMK 4, 2008.

4. Harutyunyan, Davit, et al. "On Ideal Dynamic Climbing Ropes." *Proceedings of the Institution of Mechanical Engineers, Part P: Journal of Sports Engineering and Technology* 231.2 (2017): 136-143.

Photo credits:

Laptop charger – courtesy of Ruben Post

Snow blower – courtesy of Michael Wiklund

Principle 32: Provide a handrail
References:

1. "Housing and Health Regulations in Europe." *World Health Organization*, 2017. http://www.euro.who.int/__data/assets/pdf_file/0019/121834/E89278sum.pdf.

2. "What Is UD?" *UniversalDesign.com*, Steinfeld and Maisel, 2012, www.universaldesign.com/what-is-ud/.

3. "Building Codes - Information and Guidelines." *BuyRailings.com*, 2018, www.buyrailings.com/content/generalbuildingcodes.

4. "1910.29 Fall Protection Systems and Falling Object Protection-Criteria And Practices." *United States Department of Labor*, Nov. 2016, www.osha.gov/pls/oshaweb/owadisp.show_document?p_id=9721&p_table=standards.

5. "City of Toronto Accessibility Design Guidelines." 2004. https://www.toronto.ca/wp-content/uploads/2017/08/8fcf-accessibility_design_guidelines.pdf.

Photo credits:

Airplane cabin – SSJ100 for Interjet - Interiors by SuperJet International / CC-BY-SA-2.0

Toronto skyline – Toronto at Dusk by Benson Kua / CC-BY-SA-2.0 (image modified)

Principle 33: Prevent entrapment
References:

1. Bakalar, Nicholas. "Quicksand Science: Why It Traps, How to Escape." *National Geographic News*, 28 Sept. 2005, https://news.nationalgeographic.com/news/2005/09/quicksand-science-why-it-traps-how-to-escape/

2. KidsHealth. "Choosing Safe Baby Products: Cribs." *KidsHealth.org*. The Nemours Foundation. http://kidshealth.org/en/parents/products-cribs.html.

3. "ASTM F1169–13, Standard Consumer Safety Specification for Full-Size Baby Cribs." *ASTM International*. 2013.

4. "Public Playground Safety Handbook." *U.S. Consumer Product Safety Commission (CPSC)*. Bethesda, MD: CPSC. 2008. https://www.cpsc.gov/s3fs-public/325.pdf.

5. Gibson, Kevin. "1999 - 2010 Reported Circulation/Suction Entrapments Associated with Pools, Spas and Whirlpool Bathtubs, 2011 Report." U.S. Consumer Product Safety Commission, 2011, www.cpsc.gov/s3fs-public/pdfs/entrap11.pdf.

6. The Association of Pool and Spa Professionals. "American National Standard for Suction Entrapment Avoidance in Swimming Pools, Wading Pools, Spas, Hot Tubs, and Catch Basins." *American National Standards Institute*. 11 Sept. 2006. https://www.wvdhhr.org/phs/pools/Virginia%20Graeme%20Baker%20Act/ANSI-APSP-7%202006%20suction%20entrapment%20PDF%20with%20covers.pdf.

7. Sorangel. "The Virginia Graeme Baker Pool and Spa Safety Act (VGB Act)." *Broward County & Miami Dade Pool Cleaning - Urban Pool Services*, 6 Apr. 2018, urbanpoolservices.com/virginia-graeme-baker-pool-safety/.

8. FDA. "A Guide to Bed Safety Bed Rails in Hospitals, Nursing Homes and Home Health Care: The Facts." *USFDA*. Apr. 2010. https://www.fda.gov/MedicalDevices/ProductsandMedicalProcedures/GeneralHospitalDevicesandSupplies/HospitalBeds/ucm123676.htm.

9. "Guidance for Industry and FDA Staff, Hospital Bed System Dimensional and Assessment Guidance to Reduce Entrapment." *U.S. Department of Health and Human Services, Food and Drug Administration, Center for Devices and Radiological Health*. 10 Mar. 2006. https://www.fda.gov/downloads/MedicalDevices/DeviceRegulationandGuidance/GuidanceDocuments/UCM072729.pdf.

10. Plant, Matthew A., and Jeffrey Fialkov. "Total Scalp Avulsion with Microvascular Reanastomosis: A Case Report and Literature Review." *The Canadian Journal of Plastic Surgery*, vol. 18, no. 3, 2010, pp. 112–115., www.ncbi.nlm.nih.gov/pmc/articles/PMC2940969/.

Principle 34: Encourage safe lifting

References:

1. "Lifting and Material Handling." Environment, Health and Safety, *University of North Carolina at Chapel Hill*, ehs.unc.edu/workplace-safety/ergonomics/lifting/.

2. Ergonomic Guidelines for Manual Material Handling. ser. 2007-131, *California Department of Industrial Relations*, 2007, Ergonomic Guidelines for Manual Material Handling, www.cdc.gov/niosh/docs/2007-131/pdfs/2007-131.pdf.

3. Middlesworth, Mark. "A Step-by-Step Guide to Using the NIOSH Lifting Equation for Single Tasks." *ErgoPlus*, ergo-plus.com/niosh-lifting-equation-single-task/.

Photo credits:

Patient lift – from Colourbox.com

Principle 35: Make design features congruent

References:

1. "Congruent." Google (Definition). *Google LLC*. https://www.google.com/search?q=congruent&source=lnms&sa=X&ved=0ahUKE wiW-p-akaDUAhWEShQKHdMSBGkQ_AUIBSgA&biw=1713&bih=1228&dpr=1.

2. Leung, Lily. "The 7 Principles of Conversion Centred Design." *Venture Accelerator Partners*, 6 Oct. 2015, www.vapartners.ca/7-principles-conversion-centred-design/.

3. Cherry, Kendra. "How the Stroop Effect Works." *Verywell Mind*, 10 Jan. 2018, www.verywellmind.com/what-is-the-stroop-effect-2795832.

Photo credits:

Stepladder - from Colourbox.com

Mr. Yuk – The Mr. Yuk symbol is a registered trademark oF Children's Hospital of Pittsburgh of UPMC. Used with permission.

Car seat controls – courtesy of Michael Wiklund

Principle 36: Provide visual access

References:

1. Rose, Barbara Wade. "FATAL DOSE: Radiation Deaths Linked to AECL Computer Errors." June 1994, www.ccnr.org/fatal_dose.html.

2. Nichol, Ryan J. "Airline Head-Up Display Systems: Human Factors Considerations." *International Journal of Economics & Management Sciences,* vol. 04, no. 05, 3 May 2015, doi:10.4172/2162-6359.1000248.

3. Stark, Lisa, and Enjoli Francis. "NTSB Suggests Wingtip Cameras on Planes." *ABC News,* ABC News Network, 6 Sept. 2012, abcnews.go.com/US/ntsb-suggests-wingtip-cameras-planes/story?id=17174065.

4. David, Leonard. "The Untold Story: Columbia Shuttle Disaster and Mysterious 'Day 2 Object.'" *Space.com*, 1 Feb. 2013, www.space.com/19605-columbia-shuttle-disaster-mystery-object.html.

Photo credits:

Da Vinci 3D feed – © 2018 Intuitive Surgical

Heads-Up Display – from US Air Force / US Air Force Maj. Chad E. Gibson / Public Domain

Principle 37: Make glass panes visible

References:

1. Gamet, Jeff. "Woman Sues Apple for $1M after Walking into Glass Wall." *The Mac Observer*, 26 Mar. 2012, www.macobserver.com/tmo/article/woman_sues_apple_for_1m_after_walking_into_glass_wall.

2. "Manifestation Legislation." *Windowfilm*, www.windowfilm.co.uk/graphics/manifestation-legislation.

3. "Prevent Running into Glass Doors: Glass Safety and Awareness." *Window Film World*, windowfilmworld.com/collections/clear-glass-door-safety-and-awareness-film.

4. Foderaro, Lisa W. "A City of Glass Towers, and a Hazard for Migratory Birds." *New York Times*, 14 Sept. 2011, www.nytimes.com/2011/09/15/nyregion/making-new-yorks-glass-buildings-safer-for-birds.html.

5. "The Clear Solution." *Welcome | ORNILUX Bird Protection Glass*, www.ornilux.com/.

Principle 38: Minimize distractions

References:

1. Hamilton, Jon. "Think You're Multitasking? Think Again." *NPR*, 2 Oct. 2008, www.npr.org/templates/story/story.php?storyId=95256794.

2. "NHTSA Survey Finds 660,000 Drivers Using Cell Phones or Manipulating Electronic Devices while Driving at Any Given Daylight Moment." *US Department of Transportation,* U.S. Department of Transportation's National Highway Traffic Safety Administration, 5 Apr. 2013, www.transportation.gov/briefing-room/nhtsa-survey-finds-660000-drivers-using-cell-phones-or-manipulating-electronic-devices.

3. "Traffic Safety Facts, Research Note, Distracted Driving 2014." *U.S. Department of Transportation*, U.S. Department of Transportation's National Highway Traffic Safety Administration, Apr. 2016, https://crashstats.nhtsa.dot.gov/Api/Public/ViewPublication/812260

4. "Cellular Phone Use and Texting While Driving Laws." *National Conference of State Legislatures*, 4 Apr. 2018, www.ncsl.org/research/transportation/cellular-phone-use-and-texting-while-driving-laws.aspx.

5. "How to Use the Do Not Disturb while Driving Feature." *Apple Support*, 4 May 2018, support.apple.com/en-us/HT208090.

6. "Chapter 6: Cognition, Addressing Time-Sharing Overload." *An Introduction to Human Factors Engineering*, by Christopher D. Wickens et al., Second ed., Pearson Prentice Hall, 2004, pp. 154.

7. "Simulated-Use Human Factors Validation Testing." *Food and Drug Administration*, Applying Human Factors and Usability Engineering to Medical Devices, Guidance for Industry and Food and Drug Administration Staff, 3 Feb. 2016, https://www.fda.gov/downloads/MedicalDevices/.../UCM259760.pdf

Principle 39: Prevent falls

References:

1."Most Falls Are Due to Slips and Trips on the Same Walking Surface Level." *Industrial Safety & Hygiene News*, 14 Mar. 2013, www.ishn.com/articles/95256-most-falls-are-due-to-slips-and-trips-on-the-same-walking-surface-level.

2. Abraham, Michael K, and Nicole Cimino-Fiallos. "Falls in the Elderly: Causes, Injuries, and Management." *Medscape*, 1 Feb. 2017, https://reference.medscape.com/features/slideshow/falls-in-the-elderly.

3. "CDC Features." *Centers for Disease Control and Prevention,* 22 Sept. 2017, www.cdc.gov/features/older-adult-falls/index.html.

4. "1910.140 - Personal Fall Protection Systems." *Occupational Safety and Health Administration*, United States Department of Labor, Nov. 2016, www.osha.gov/laws-regs/regulations/standardnumber/1910/1910.140.

5. Ridenour, Marcella V. "Age, Side Height, and Spindle Shape of the Crib In Climbing over the Side." *Perceptual and Motor Skills,* vol. 85, no. 2, Oct. 1997, pp. 667–674., doi:10.2466/pms.85.6.667-674.

Photo credits:

Man's hands – from Pixnio / Public Domain

Workers on mast – from Pxhere / Public Domain

Principle 40: Make it slip resistant
References:

1. "CDC Features." *Centers for Disease Control and Prevention*, 22 Sept. 2017, www.cdc.gov/features/older-adult-falls/index.html.

2. "Important Facts about Falls." *Centers for Disease Control and Prevention*, 10 Feb. 2017, www.cdc.gov/homeandrecreationalsafety/falls/adultfalls.html.

3. Bruzek, Joe. "2015 Ford Mustang: The Pros and Cons of Optional Recaro Seats." *Cars.com*, 1 Oct. 2016, www.cars.com/articles/2015-ford-mustang-the-pros-and-cons-of-optional-recaro-seats-1420681061039/.

4. "Tribometer." *Revolvy.*, www.revolvy.com/main/index.php?s=Tribometer&item_type=topic.

5. "ISO 25178-1:2016(En): Geometrical product specifications (GPS) — Surface texture: Areal — Part 1: Indication of Surface Texture." *ISO - International Organization for Standardization*, www.iso.org/obp/ui#iso:std:iso:25178:-1:ed-1:v1:en:sec:foreword.

6. "What Is a Friction Burn?" *Sharecare*, www.sharecare.com/health/burns/what-is-a-friction-burn.

Photo credits:

Bicycle handle – courtesy of Brenda van Geel

Car seat – from user: PFWOuoh / Wikimedia Commons / CC-BY-SA-4.0

Dental tool – from user: AfroBrazilian / Wikimedia Commons / CC-BY-SA-3.0

Handrail – courtesy of Valerie Ng

Motorcycle crash – from user: driver Photographer / Flickr / CC-BY-SA-2.0

Exemplar 4: Stepladder
References:

1. "Portable Ladder Safety." *Occupational Safety and Health Administration, United States Department of Labor*, www.osha.gov/Publications/portable_ladder_qc.html.

Person on stepladder – courtesy of Rachel Aronchick

Principle 41: Display critical information continuously
References:

1. Administration, Federal Aviation. *Helicopter Flying Handbook: FAA-H-8083-21a*. United States Department of Transportation, Federal Aviation Administration, Airman Testing Standards Branch, 2012.

2. Clay. "Airplane Instrument Basics: The Six Pack." *Clayviation*, 11 Jan. 2017, clayviation.com/2017/01/11/airplane-instrument-basics-the-six-pack/.

3. Andrade, Mário R. "The Gutenberg Diagram in Web Design – User Experience – Medium." *Medium, User Experience*, 9 June 2013, medium.com/user-experience-3/the-gutenberg-diagram-in-web-design-e5347c172627.

4. Design Principles: Visual Perception and the Principles of Gestalt." *Smashing Magazine,* 28 Mar. 2014, www.smashingmagazine.com/2014/03/design-principles-visual-perception-and-the-principles-of-gestalt/.

5. "Banner Blindness Revisited: Users Dodge Ads on Mobile and Desktop." *Nielsen Norman Group*, www.nngroup.com/articles/banner-blindness-old-and-new-findings/.

Photo credits:

Airplane dashboard – from Oscar Sutton / Unsplash

Principle 42: Prevent users from disabling alarms
References:

1. O'Brien, Dan. "Audible Alarm Basics | Everything You Wanted to Know, but Were Afraid to Ask." *Digi-Key Electronics*, www.digikey.com/en/pdf/m/mallory-sonalert-products/mallorysonalert-audiblealarmbasics/.

2. "What Sounds Can People Hear?" *Discovery of Sound in the Sea*, University of Rhode Island and Inner Space Center, 2017, dosits.org/science/measurement/what-sounds-can-we-hear/.

3. Raine, Jordan. "Why It's So Hard to Ignore a Baby's Cry, According to Science." *The Conversation*, 30 Aug. 2016, theconversation.com/why-its-so-hard-to-ignore-a-babys-cry-according-to-science-63245.

4. "Sonic Science: The High-Frequency Hearing Test." *Scientific American*, 23 May 2013, www.scientificamerican.com/article/bring-science-home-high-frequency-hearing/.

5. "Alaris™ System with Guardrails™ Suite MX." *CareFusion*, Dec. 2016. https://www.bd.com/documents/guides/user-guides/IF_Alaris-System-8015-v9-19_UG_EN.pdf.

6. "Audible Visible Appliance Reference Guide." Applications Guide. *System sensor*. https://www.firelite.com/CatalogDocuments/SSD_AV_Application_Guide.pdf.

7. "Fire Alarm Concerns." *Duke Occupational & Environmental Safety Office*, www.safety.duke.edu/sites/default/files/Fire%20Alarm%20Concerns.pdf.

8. Wilder, Rob. "8 Ways to Reduce Alarm Fatigue in Hospitals." *Spok*, 19 Sept. 2017, www.spok.com/blog/8-ways-reduce-alarm-fatigue-hospitals.

Photo credits:

Baby monitor – courtesy of Erin Davis

Fire alarm – Ben Schumin / CC-BY-SA-2.0

Infusion pump – from Colourbox.com

Principle 43: Provide undo option

References:

1. Barnet, Belinda. "Crafting the User-Centered Document Interface: The Hypertext Editing System (HES) and the File Retrieval and Editing System (FRESS)." *Digital Humanities Quarterly*, vol. 4, no. 1, 2010, www.digitalhumanities.org/dhq/vol/4/1/000081/000081.html#vandam1999.

2. Nessif, Bruna. "Don't Worry, Tinder Users! You Will Soon Be Able to "Undo" That Left Swipe, but It'll Cost You." *E! News*, 5 Nov. 2014, www.eonline.com/news/595269/don-t-worry-tinder-users-you-will-soonbe-able-to-undo-that-left-swipe-but-it-ll-cost-you.

Principle 44: Avoid toggle ambiguity

1. Nguyen, Kevin. "The Little Switch." *The Atlantic*, 5 Nov. 2013, www.theatlantic.com/technology/archive/2013/11/the-little-switch/281041/.

Photo credits:

Automotive toggle switch – from user: The Car Spy / Flickr / CC-BY-2.0

Power switch, toggle switch physical guard – courtesy of Kimmy Ansems

Principle 45: Require professional maintenance and repair

References:

1. Trigg, Andy. "Shock Warning If Repairing an Appliance That's Turned Off." *White Goods Help and Advice for UK*, 24 May 2017, www.whitegoodshelp.co.uk/can-you-still-get-a-shock-repairing-an-appliance-if-its-turned-off/.

Photo credits:

Battery – from Max Pixel / Public Domain

Dialysis machine – used with permission from Baxter

Principle 46: Use telephone-style keypad layout

References:

1. Campbell, Todd. "Answer Geek: Calculator vs. Phone Layouts." *ABC News*, ABC News Network, abcnews.go.com/Technology/story?id=119296&page=1.

2. "Phone Key Pads." *Dial ABC*, www.dialabc.com/motion/keypads.html.

3. Smeltz, Adam. "50 Years Ago, Touch-Tone Phones Began a Communication Revolution." *TribLIVE.com*, 17 Nov. 2013, triblive.com/news/allegheny/4291278-74/phone-touch-tone.

4. Shneiderman, Ben, et al. Designing the User Interface: Strategies for Effective Human-Computer Interaction. Pearson, 2010.

5. NHS, National Patient Safety Agency. "Design For Patient Safety: A Guide to the Design of Electronic Infusion Devices." 2010, www.nrls.npsa.nhs.uk/EasySiteWeb/getresource.axd?AssetID=68536.

<u>Photo credits:</u>

Touch tone phone – from Wikimedia Commons / Public Domain

Principle 47: Indicate unsaved changes
<u>References:</u>

1."Pictures of Airbus A-320's Ditch Switch: Update on Flight 1549." *Popular Mechanics*, 15 Feb. 2018, www.popularmechanics.com/flight/a12346/4299756/.

Principle 48: Make text legible
<u>References:</u>

1. "6.2.2.5 Visual angle" and "6.2.2.6.5 Minimum type size". *ANSI/AAMI HE 75:2009, Human factors engineering: Design of medical devices.* Association for the Advancement of Medical Instrumentation, 2010, pp. 46, 48-49.

2. "19.4.1.2 Optimal character height". *ANSI/AAMI HE 75:2009, Human factors engineering: Design of medical devices.* Association for the Advancement of Medical Instrumentation, 2010, pp. 299.

3. "Character and Symbol Size." *FAA Human Factors.* US Federal Aviation Administration. http://www.hf.faa.gov/Webtraining/VisualDisplays/text/size1a.htm.

4. "1.4.3 Contrast (Minimum)" and "1.4.6 Contrast (Enhanced)." *W3C Web Content Accessibility Guidelines (WCAG) 2.0.* World Wide Web Consortium, (MIT, ERCIM, Keio, Beihang), 2008. https://www.w3.org/TR/WCAG20/.

5. (For illustrative purposes only.) Contrast in print might not match values as measured in digital form using Contrast Analyser, available at http://www.paciellogroup.com/resources/contrast-analyser.html.

Principle 49: Provide backup display
<u>References:</u>

1. Nagelhout, John J., and Karen L. Plaus. *Nurse Anesthesia.* 5th ed., Elsevier Saunders, 2014.

2. "Electronic Flight Instruments." *Advanced Avionics Handbook*, FAA-H-8083-6, Federal Aviation Administration, www.faa.gov/regulations_policies/handbooks_manuals/aviation/advanced_avionics_handbook/media/aah_ch02.pdf.

<u>Photo credits:</u>

Cockpit – from user: Naddsy / Flickr / CC-BY-SA-2.0

Dead pixels – from user: Kprateek88 / Wikimedia Commons / CC-BY-SA

Screen hardware failure – from Sibe Kokke / Flickr / CC-BY-SA-2.0

Screen software failure – from Taber Andrew Bain / Flickr / CC-BY-SA-2.0

Principle 50: Use color-coding
<u>References:</u>

1. "Why Does Red Mean Stop and Green Mean Go?" *Mental Floss*, 7 July 2014, www.mentalfloss.com/article/57267/why-does-red-mean-stop-and-green-mean-go.

2. Cutolo, Morgan. "Why Are Traffic Lights Red, Yellow, and Green?" *Reader's Digest,* 26 May 2017, www.rd.com/advice/travel/traffic-lights/.

3. Calvert, J.B. "Early Railway Signals," mysite.du.edu/~etuttle/rail/sigs.htm.

4. Bravo, Mary J., and Ken Nakayama. "The Role of Attention in Different Visual-Search Tasks." *Perception & Psychophysics*, vol. 51, no. 5, 1992, pp. 465–472., doi:10.3758/bf03211642.

5. Peckham, Geoffrey. "Designing Effective Product Safety Labels: How to Convey Risk Severity Levels." *InCompliance*, October 2014. https://incompliancemag.com/article/designing-effective-product-safety-labels-how-to-convey-risk-severity-levels/.

<u>Photo credits:</u>

Escalator emergency stop – courtesy of Michael Wiklund

Exemplar 5: Anesthesia machine

<u>Photo credits:</u>

Principle 51: Enable emergency shutdown

<u>References:</u>

1. "Emergency Stop Switches. A Technical Guide for Proper Selection." http://eao.com/fileadmin/documents/PDFs/en/08_whitepapers/EAO_WP_Emergency-stop-switches-guide_EN.pdf (2014-06-17).

2. "IEC 60417 — Graphical Symbols for Use on Equipment." *International Organization for Standardization*, www.iso.org/obp/ui#iec:grs:60417:5638.

3. *IEC 60947-5-5 Low-Voltage Switchgear and Controlgear. Electrical Emergency Stop Device with Mechanical Latching Function.* International Electrotechnical Commission, 1997

4. "Guide to Application of the Machinery Directive 2006/42/EC." Edited by Ian Fraser, June 2010, ec.europa.eu/DocsRoom/documents/9483/attachments/1/translations/en/renditions/native.

<u>Photo credits:</u>

Principle 52: Shield or isolate from heat

<u>References:</u>

1. "Pain Caused by Burns." Edited by Carol DerSarkissian, *WebMD*, 30 Apr. 2017, www.webmd.com/pain-management/guide/pain-caused-by-burns.

2. Parsons, Ken. *Human Thermal Environments.* 2nd ed., Taylor & Francis, 2003.

3. "What Is the Temperature of a 100 Watt Bulb?" *Pacific Lamp & Supply Company*, www.pacificlamp.com/temperature-of-a-100-watt-bulb.asp.

<u>Photo credits:</u>

Principle 53: Install a physical shield

<u>References:</u>

1. Kim, Sang Min, et al. "The Relation Between the Amount of Sunscreen Applied and the Sun Protection Factor In Asian Skin," vol. 62, no. 2, 7 Dec. 2009, pp. 218–222., doi:https://doi.org/10.1016/j.jaad.2009.06.047.

2. "What Type and Thickness of Bullet Resistant Glass Do You Need?" *Creative Industries, Inc.*, 28 June 2017, cibulletproof.com/type-thickness-bullet-resistant-glass-need/.

3. Angstrom Tech Admin. "What's the Difference Between Positive and Negative Air Pressure Cleanrooms?" *Angstrom Technology,* 20 Apr. 2016, angstromtechnology.com/whats-the-difference-between-positive-and-negative-air-pressure-cleanrooms/.

4. Fenical, Gary. "The Basic Principles of Shielding." *In Compliance Magazine*, 1 Mar. 2014, incompliancemag.com/article/the-

basic-principles-of-shielding/.

<u>Photo credits:</u>

Grinder, fan – from Colourbox.com

Hard hat, helmet and pads – courtesy of Cory Costantino

Principle 54: Make blades very sharp

<u>References:</u>

1. Holloway, Beth. "Be Careful with Kitchen Knives." *University of Rochester Medical Center*, www.urmc.rochester.edu/encyclopedia/content.aspx?contenttypeid=1&contentid=263.

2. Török, Ervin. *Surgery of the Eye: A Hand-book for Students and Practitioners*. Lea & Febiger, 1913.

3. P, Sheri. Calphalon Classic Self-Sharpening 12-Pc. Cutlery Set. *Calphalon*, 1 Jan. 2018, www.calphalon.com/en-US/calphalon-classic-self-sharpening-12-pc-cutlery-set-ca-1924555--1.

4. "CS1500 Self-Sharpening Electric Chain Saw." *Oregon*, en.oregonproducts.com/pro/products/corded/CS1500.htm.

5. Walker, Stephen, and Pukko. "What's the Best Steel for a Kitchen Knife?" *KnifePlanet.net*, 12 Apr. 2017, www.knifeplanet.net/best-steel-kitchen-knife/.

6 "More Advice and Theory on Sharpening Angles for Knives." *SharpeningSupplies.com*, www.sharpeningsupplies.com/Detailed-Discussion-on-Knife-Sharpening-Angles-W28.aspx.

<u>Photo credits:</u>

Incision – from Senior Airman John Nieves Camacho, U.S. Air Force / Public Domain

Man sharpening knife – from user: Didriks / Flickr / CC-BY-SA-2.0

Sharpening block – used with permission from Calphalon Corp

Principle 55: Start on slow and low

<u>References:</u>

1. Mayo Clinic Staff. "CPAP Machines: Tips for Avoiding 10 Common Problems." *Mayo Clinic*, Mayo Foundation for Medical Education and Research, 17 May 2018, www.mayoclinic.org/diseases-conditions/sleep-apnea/in-depth/cpap/art-20044164.

2. "How Titration May Help Your Child with ADHD." *WebMD*, 2016, www.webmd.com/add-adhd/titration-treatment-adhd#1. Reviewed by Smitha Bhandari, MD on August 21, 2016.

3. "Program for Psychiatric Drug Withdrawal." *Drug Withdrawal Research Foundation | Psychiatric Medication Tapering Program*, Drug Withdrawal Research Foundation, 2017, www.withdrawalresearch.org/tapering-program.html.

4. Shoreline Aviation, Inc. "Transitioning to Jet Flight." *Shoreline Aviation*, Shoreline Aviation, 22 Aug. 2014, www.shorelineaviation.net/news---events/bid/72339/Transitioning-to-Jet-Flight.

Principle 56: Use sensors

<u>References:</u>

1. Kenney, Briley. "Top 10 Wearables for Your Baby for Peace of Mind." *SmartWatches.org*, 25 Feb. 2016, smartwatches.org/fitness/top-10-wearables-for-your-baby./

2. Smouse, Becca. "New Car Tech Could Stop Drunken Drivers." *USA Today*, Gannett Satellite Information Network, 7 July 2015, www.usatoday.com/story/money/cars/2015/07/06/new-technology-to-prevent-drunk-driving/29125417.

3. Walsh, Michael. "Alcohol Detecting Technology Could Save 10,000 a Year from Drunk-Driving Death: Scientists." *Nydailynews.com*, *New York Daily News*, 4 Jan. 2013, www.nydailynews.com/news/national/alcohol-detecting-technology-save-10-000-year-drunk-driving-death-article-1.1231763.

4. "How Does SawStop Work?", *SawStop*, SawStop, www.sawstop.com/why-sawstop/the-technology./

5. Ahrens, Marty. "Smoke Alarms in U.S. Home Fires." *National Fire Protection Association*, NFPA.org, Sept. 2015, www.nfpa.org/News-and-Research/Data-research-and-tools/Detection-and-Signaling/Smoke-Alarms-in-US-Home-Fires.

6. Wilson, Chauncey, and Nigel Bevan . "Usability Body of Knowledge." *Usability Body of Knowledge*, User Experience Professionals' Association, June 2009, www.usabilitybok.org/function-allocation.

7. "Table 2 - Human versus Machine Capabilities." *IEC TR 62366-2 | Medical Devices – Part 2: Guidance on the Application of Usability Engineering to Medical Devices*, 1.0 ed., International Electromechanical Commission, 2016.

8. Sendelbach, Sue, and Marjorie Funk. "Alarm Fatigue Symposium | Patient Safety Issues in Critical Care | A Patient Safety Concern." *Aacn.org*, American Association of Critical-Care Nurses, www.aacn.org/education/publications/acc/24/4/0378-symposium-patient-safety-issues-in-critical-care-alarm-fatigue-a-patient-safety-concern.

9. Linthicum, David. "When Sensors Don't Make Sense: Ridiculous IoT Devices - CTP." *Cloud Technology Partners*, 1 June 2016, www.cloudtp.com/doppler/sensors-dont-make-sense-ridiculous-iot-devices/.

Photo credits:

Baby monitor – courtesy of Allison Strochlic

Alcohol sensing car, fire alarm – courtesy of Cory Costantino

SawStop – used with permission from SawStop

Principle 57: Enable escape

References:

1. Department of Transportation, Federal Aviation Administration. "14 CFR Parts 25 and 121 Revision of Emergency Evacuation Demonstration Procedures to Improve Participant Safety; Final Rule." Final Rule, 2004.https://www.gpo.gov/fdsys/pkg/FR-2004-11-17/pdf/04-25493.pdf

2. "Safety Information Card: Silk Air, Airbus A320 | San Francisco International Airport." *SFO Museum*, www.flysfo.com/museum/aviation-museum-library/collection/18548.

3. Steve. "How a Refrigerator Can Kill You." *The Refrigerator and Fridge Freezer Site*, 17 Mar. 2011, www.fridgefreezersite.com/a-refrigerator-can-kill-you/.

4. Hester, Donald. "Emergency Egress and Rescue Openings per the 2015 IRC." *Wenatchee Home Inspections | NCW Home Inspections, LLC, InterNACHI*, 22 Jan. 2017, www.ncwhomeinspections.com/Emergency+Egress+and+Rescue+openings+per+the+2015+IRC.

5. "Federal Motor Vehicle Safety Standards; Interior Trunk Release." *Federal Register*, 20 Oct. 2000, www.federalregister.gov/documents/2000/10/20/00-27038/federal-motor-vehicle-safety-standards-interior-trunk-release.

6. Geiger, Jennifer. "Emergency Trunk-Release Lever Saves Lives | News from Cars.com." *Cars.com*, 21 Sept. 2012, www.cars.com/articles/2012/09/emergency-trunk-release-lever-saves-lives/.

Principle 58: Detect fatigue and rouse users

References:

1. "Research on Drowsy Driving." *National Highway Traffic Safety Administration,* US Department of Transportation, one.nhtsa.gov/Driving-Safety/Drowsy-Driving/scope–of–the–problem.

2. "Exeros Technologies." *Sleep Watcher XR - Driver Fatigue Alarm, Driver Alert System*, www.exeros-technologies.com/products/driver-fatigue-management.

3. Duquette, Alison. "Fact Sheet – Pilot Fatigue Rule Comparison." *United States Department of Transportation*, Federal Aviation Administration, 21 Dec. 2011, www.faa.gov/news/fact_sheets/news_story.cfm?newsId=13273.

4. McCormick, Frank, et al. "Surgeon Fatigue." *Archives of Surgery*, vol. 147, no. 5, May 2012, pp. 430–435., doi:10.1001/archsurg.2012.84.

5. "Drunk Driving." *United States Department of Transportation*, NHTSA, www.nhtsa.gov/risky-driving/drunk-driving.

6. Taub, Eric A. "Sleepy behind the Wheel? Some Cars Can Tell." Wheels, *The New York Times*, 16 Mar. 2017, www.nytimes.com/2017/03/16/automobiles/wheels/drowsy-driving-technology.html.

7. Marks, Paul. "Drowsiness Detector Wakes Drivers If They Start to Doze." New Scientist, *Daily News*, 28 May 2013, www.newscientist.com/article/dn23604-drowsiness-detector-wakes-drivers-if-they-start-to-doze/.

8. Shurkin, Joel. "Watching the Car and Shaking the Wheel to Wake Sleepy Drivers." *Inside Science*, 2 July 2015, 18:45, www.insidescience.org/news/watching-car-and-shaking-wheel-wake-sleepy-drivers.

9. Kozak, Ksenia, et al. "Evaluation of Lane Departure Warnings for Drowsy Drivers," *Proceedings of the Human Factors and Ergonomics Society 50th Annual Meeting*. 2006, pp. 2400–2404. https://pdfs.semanticscholar.org/1d21/ae8564bb774aa0f255366cebfc86a099a829.pdf.

Photo credits:

Car interior – from Pexels

FAA logo – from Wikimedia Commons / Public Domain

Principle 59: Augment control

References:

1. "From Innovation to Standard Equipment - 30 Years of Safe Braking with Bosch ABS. "*Automotive Technology*, https://www.bosch.co.jp/en/press/group-0807-05.asp.

2. United States, Congress, Government Publishing Office. "49 CFR 393.55. Antilock Brake Systems." 2011, pp. 935-936. https://www.gpo.gov/fdsys/pkg/CFR-1999-title49-vol4/pdf/CFR-1999-title49-vol4-sec393-55.pdf.

3. Goreham, John. "Buyer's Guide: Braking Technology – What's Standard, and Which Options to Choose." *BestRide.com*, bestride.com/research/buyers-guide/braking-technology-2016.

4. Pope, Stephen. "Fly by Wire: Fact versus Science Fiction." *Flying Magazine*, 23 Apr. 2014, www.flyingmag.com/aircraft/jets/fly-by-wire-fact-versus-science-fiction.

5. Alltech. *MEGA Range Limiting Device*, www.groupealltech.com/EN/products/mega.

6. "Bagel Danger." *Freakonomics Blog*, 30 Nov. 2009, freakonomics.com/2009/11/30/bagel-danger/

Photo credits:

Bagel slicer – courtesy of Rachel Aronchick

Principle 60: Reduce (or isolate from) vibration

References:

1. Shen, Shixin (Cindy), and Ronald A. House. "Hand-Arm Vibration Syndrome." *Canadian Family Physician*, vol. 63, no. 3, Mar. 2017, pp. 206–210., www.ncbi.nlm.nih.gov/pmc/articles/PMC5349719/.

2. Morrison, Kyle W. "Whole-Body Vibration." *Safety + Health*. 1 Oct. 2009. National Safety Council. http://www.safetyandhealthmagazine.com/articles/whole-body-vibration-2.

3. "Vibration Syndrome." *Centers for Disease Control and Prevention*. DHHS (NIOSH) Publication Number 83-110. Mar 1983. https://www.cdc.gov/niosh/docs/83-110/default.html.

4. "Directive 2002/44/Ec of the European Parliament and of the Council of 25 June 2002." The European Parliament and the Council of the European Union, 25 June 2002. eur-lex.europa.eu/legal-content/EN/TXT/PDF/?uri=CELEX:02002L0044-20081211&from=EN.

5. "Human Vibration." *Bruel & Kjaer*. Nov 1989. https://www.bksv.com/~/media/literature/Primers/br056.ashx?la=en.

6. Bob, Cheung, and Nakashima Ann. A Review on the Effects of Frequency of Oscillation on Motion Sickness. *Defence R&D Canada Toronto*. 2006. https://apps.dtic.mil/dtic/tr/fulltext/u2/a472991.pdf

7. Du, Bronson Boi, et al. "The Impact of Different Seats and Whole-Body Vibration Exposures on Truck Driver Vigilance and Discomfort." Ergonomics, vol. 61, no. 4, 2017, pp. 528–537., doi:10.1080/00140139.2017.1372638.

8. Dong, Ren G. et al. "Tool-Specific Performance of Vibration-Reducing Gloves for Attenuating Palm-Transmitted Vibrations in Three Orthogonal Directions." *International Journal of Industrial Ergonomics*. 44.6 (2014): 827–839. PMC.

9. "Densimet® Tool Holders." *Plansee Group*. https://www.plansee.com/en/products/components/forming-andmachining-tools/vibration-damping-tool-holders.html.

10. "The Art of Ergonomics." *Atlas Copco Industrial Technique AB*. 2001. https://www.atlascopco.com/content/dam/atlas-copco/industrial-technique/general/documents/pocketguides/Pocket%20guide_the%20art%20of%20ergonomics.pdf.

Photo credits:

Simulated white-tip syndrome – courtesy of Cory Costantino

Worker with grinder – from Colourbox.com

Exemplar 6: Chainsaw

Photo credits:

Woodcutter – from Pixabay / Public Domain

Principle 61: Prevent fluid ingress

References:

1. Bisenius, William S. *WebCite Query Result*, 29 Dec. 2012, www.webcitation.org/6DGYoRMwp?url=http%3A%2F%2Fwww.ce-mag.com%2Farchive%2F06%2FARG%2Fbisenius.htm.

Photo credits:

Gas nozzle – from Pixabay / Public Domain

Phone gasket – courtesy of Kimmy Ansems

Principle 62: Use "TALLman" lettering

References:

1. "FDA and ISMP Lists of Look-Alike Drug Names with Recommended Tall Man Letters." *Institute for Safe Medication Practices*, 2016, https://www.ismp.org/tools/tallmanletters.pdf.

2. "Use of Tall Man Letters Is Gaining Wide Acceptance." *Institute For Safe Medication Practices*, 31 July 2008, https://www.ismp.org/resources/use-tall-man-letters-gaining-wide-acceptance.

3. "Name Differentiation Project." *US Food and Drug Administration Home Page*, Center for Drug Evaluation and Research, 6 Nov. 2017, www.fda.gov/Drugs/DrugSafety/MedicationErrors/ucm164587.htm.

4. "Special Edition: Tall Man Lettering; ISMP Updates Its List of Drug Names with Tall Man Letters. *Institute For Safe Medication Practices*, 2 June 2016, https://www.ismp.org/newsletters/acutecare/showarticle.aspx?id=1140.

Photo credits:

Medication bottles – Photo courtesy of Institute for Safe Medication Practices

Spectrum IQ Infusion Pump – used with permission from Baxter International Inc.

Principle 63: Childproof hazardous items

References:

1. Sengölge, M. and Vincenten, J. "Child Product Safety Guide: Potentially Dangerous Products." *European Child Safety Alliance*, EuroSafe, Nov. 2013, doi:10.3403/30311725, http://www.childsafetyeurope.org/publications/info/product-safety-guide.pdf.

2. Robinson, Abby. "Georgia Institute of Technology." *Georgia Tech News Center*, 5 Nov. 2013, www.news.gatech.edu/2013/11/02/researchers-help-design-easy-open-child-resistant-medicine-bottle.

3. "Children Can Open Medicines with an 'Arthritis Cap.'" *Consumermedsafety.org*, 28 Feb. 2014, 16:17, www.consumermedsafety.org/safe-medicine-storage-and-disposal/children-can-open-medicines-with-an-arthritis-cap.

4. Poison Prevention Packaging Act." *CPSC.gov*, 9 Jan. 2018, www.cpsc.gov/Regulations-Laws--Standards/Statutes/Poison-Prevention-Packaging-Act.

5. "16 CFR 1700.20 - TESTING PROCEDURE FOR SPECIAL PACKAGING." The U.S. Government Publishing Office, Jan. 2012, doi:10.1515/9783110818840-008, https://www.gpo.gov/fdsys/pkg/CFR-2012-title16-vol2/pdf/CFR-2012-title16-vol2-sec1700-20.pdf.

Photo credits:

Bumper, cabinet latch, plug protector, stair gate – courtesy of Kimmy Ansems

Child with toy – from Pixabay / Public Domain

High chair – from Colourbox.com

Principle 64: Indicate expiration date

References:

1. Leib, Emily B. et al. "The Dating Game: How Confusing Food Date Labels Lead to Food Waste in America." *Harvard Food Law and Policy Clinic and the Natural Resources Defense Council*, September 2013. http://www.chlpi.org/wp-content/uploads/2013/12/dating-game-report.pdf.

2. Skinner, Ginger. "What You Need to Know about Expired EpiPens: Are They Safe? And Will They Be Effective in an Emergency?" *Consumer Reports*, 26 Aug 2016. http://www.consumerreports.org/drugs/expired-epipens-what-you-need-to-know/.

3. "Smoke Alarm Outreach Materials." *U.S. Fire Administration*, 29 Mar. 2018, www.usfa.fema.gov/prevention/outreach/smoke_alarms.html#ans5.

4. Center for Food Safety and Applied Nutrition. "Shelf Life/Expiration Dating." *U.S. Food and Drug Administration*, www.fda.gov/Cosmetics/Labeling/ExpirationDating/default.htm.

5. Tedesco, Laura. "How Germy Is Your Makeup?" *Women's Health*. 23 August 2013. http://www.womenshealthmag.com/health/how-germy-is-your-makeup.

Principle 65: Make packages easy to open

References:

1. Latimer, Cole. "Four out of Five People Suffer Package-Rage Due to Poor Packaging, Survey Reveals." *Ferret*, 12 Nov. 2013, www.ferret.com.au/articles/in-focus/four-out-of-five-people-suffer-package-rage-due-to-poor-packaging-survey-reveals-n2510193.

2. Coffin, Katie. "Securing Success: Allegion Thrives on Strong Values." *BizVoice/Indiana Chamber*, 2015, pp. 73, 78., www.bizvoicemagazine.com/wp-content/uploads/2018/03/Allegion.pdf.

3. Peterson, Josh. "Thousands Injured by 'Diabolical' Packaging." *TreeHugger*, 18 Jan. 2009, www.treehugger.com/green-food/thousands-injured-by-adiabolicala-packaging.html.

4. "Amazon Certified Frustration-Free Packaging." *Amazon*, www.amazon.com/b/?&node=5521637011.

Principle 66: Fight bad bacteria

References:

1. Sender, Ron, et al. "Revised Estimates for the Number of Human and Bacteria Cells in the Body." *PLOS Biology*, vol. 14, no. 8, 19 Aug. 2016, doi:10.1371/journal.pbio.1002533.

2. "Gut Bacteria That 'Talk' to Human Cells May Lead to New Treatments." *ScienceDaily*, Rockefeller University, 30 Aug. 2017, www.sciencedaily.com/releases/2017/08/170830141248.htm.

3. "Bacteria and Viruses." *FoodSafety.gov*, U.S. Department of Health and Human Services, www.foodsafety.gov/poisoning/causes/bacteriaviruses/.

4. "Antibiotic / Antimicrobial Resistance." *Centers for Disease Control and Prevention*, 10 Apr. 2017, www.cdc.gov/drugresistance/federal-engagement-in-ar/national-strategy/index.html.

5. Tavernise, Sabrina. "Antibiotic-Resistant Infections Lead to 23,000 Deaths a Year, C.D.C. Finds." *New York Times*, 16 Sept. 2013, www.nytimes.com/2013/09/17/health/cdc-report-finds-23000-deaths-a-year-from-antibiotic-resistant-infections.html.

6. "Handwashing: Clean Hands Save Lives." *Centers for Disease Control and Prevention*, 7 Mar. 2016, www.cdc.gov/handwashing/when-how-handwashing.html.

7. "EPA Registration." *Antimicrobial Copper*, International Copper Association, 2015, www.antimicrobialcopper.org/us/epa-registration.

8. "How It Works." *Antimicrobial Copper*, International Copper Association , 2015, www.antimicrobialcopper.org/us/how-it-works.

9. "It's Time to Redefine What Paint Can Do." *Sherwin-Williams*, 2018, www.swpaintshield.com/.

10. Intagliata, Christopher. "B.O. Gives Up Its Stinky Secrets." *Scientific American*, 3 Apr. 2015, www.scientificamerican.com/podcast/episode/b-o-gives-up-its-stinky-secrets/.

Anti-bacterial lotion dispenser, and deodorant – courtesy of Cory Costantino

Paint roller – courtesy of Cory Costantino and May Coppers Costantino

Copper cart handle – Copper Development Association, Inc., used with permission from Kellen Inc.

Principle 67: Number instructional text

<u>Photo credits:</u>

Syringe – from Pixabay / Public Domain

Principle 68: Prevent and expose tampering

<u>References:</u>

1. Markel, Howard. "How the Tylenol Murders of 1982 Changed the Way We Consume Medication." *PBS*, 29 Sept. 2014, www.pbs.org/newshour/health/tylenol-murders-1982.

2. United States, Congress, Bills and Statutes. "18 U.S.C. 1365 - Tampering with Consumer Products." 18 U.S.C. 1365 - Tampering with consumer products, U.S. Government Publishing Office, 3 Jan. 2012. www.gpo.gov/fdsys/pkg/USCODE-2011-title18/pdf/USCODE-2011-title18-partI-chap65-sec1365.pdf.

3. "Tampering." *Merriam-Webster*, www.merriam-webster.com/dictionary/tampering.

4. Gale, Thomas. "Product Tampering." *Encylopedia.com*, World of Forensic Science, 2005, www.encyclopedia.com/history/united-states-and-canada/us-history/product-tampering.

<u>Photo credits:</u>

Blister pack – from Pixabay / Public Domain

Bottle seal – from Free Stock Photos.biz / Public Domain

Heat shrink seal, safety strip – courtesy of Valerie Ng

Principle 69: Put a cap on it

<u>References:</u>

1. Perry, Jane, et al. "Scalpel Blades: Reducing Injury Risk." *Advances in Exposure Prevention*, vol. 6, no. 4, 2003, pp. 37–40., www.sandelmedical.com/management/uploads/ScalpelBladeRisk.pdf.

2. Occupational Safety and Health Administration. Concrete and Masonry Construction. 29 CFR 1926.701(b), U.S. Department of Labor, Occupational Safety and Health Administration, (OSHA), 1926.

3. Bekele, Tolesa, et al. "Factors Associated with Occupational Needle Stick and Sharps Injuries among Hospital Healthcare Workers in Bale Zone, Southeast Ethiopia." *PLOS ONE, Public Library of Science*, 15 Oct. 2015, https://doi.org/10.1371/journal.pone.0140382.

4. "Needlestick Injuries." *WHO, World Health Organization*, www.who.int/occupational_health/topics/needinjuries/en/.

5. Merhige, John A., and Lisa Caparra. "The Credence Companion Syringe System Delivers on Safety & Usability Using Human Factors Studies." *Drug Development & Delivery*, vol. 15, no. 8, Oct. 2015, www.credencemed.com/wp-content/uploads/2014/08/DDD-October-2015-Credence-Companion-and-HF-Studies-Low-Res.pdf.

6. Occupational Safety and Health Administration. "Bloodborne pathogens." 29 CFR 1910.1030(d)(2)(vii)(A), U.S. Department of Labor, Occupational Safety and Health Administration, (OSHA), 1910.

7. "Module 4: Preventing Needlestick Injuries." www.path.org/publications/files/SafeInjPDF-Module4.pdf.

8. Center for Devices and Radiological Health. "Safely Using Sharps (Needles and Syringes) at Home, at Work and on Travel: What to Do If You Can't Find a Sharps Disposal Container." *U.S. Food and Drug Administration*, www.fda.gov/MedicalDevices/ProductsandMedicalProcedures/HomeHealthandConsumer/ConsumerProducts/Sharps/ucm263259.htm.

<u>Photo credits:</u>

Injection – from US Military Health / Public Domain

Integrated syringe cap, knife, scalpel and syringe – courtesy of Valerie Ng

Steel rod end caps – from Colourbox.com

Principle 70: Use hypoallergenic material

References:

1. Nutten, Sophie. "Atopic Dermatitis: Global Epidemiology and Risk Factors." *Annals of Nutrition and Metabolism*, Karger Publishers, 24 Apr. 2015, www.karger.com/Article/FullText/370220.

2. "Metal Hypersensitivity." *MedBroadcast.com*, www.medbroadcast.com/condition/getcondition/metal-hypersensitivity.

3. "Latex Allergy." *ACAAI Public Website,* acaai.org/allergies/types/latex-allergy.

4. "Implantable Cardioverter-Defibrillators (ICDs)." *Implantable Cardioverter-Defibrillators (ICDs),* Johns Hopkins Medicine, www.hopkinsmedicine.org/heart_vascular_institute/conditions_treatments/treatments/implantable_cardioverter_defribrillators.html.

Photo credits:

Ceramic braces – from Colourbox.com

Fiberglass, gloves – courtesy of Kimmy Ansems

Hat – from Mammut, PPR/photographer

ICD device – Reproduced with permission of Medtronic, Inc.

Exemplar 7: Medication blister pack

Photo credits:

Blister packs – from Pixabay / Public Domain

Principle 71: Manage and stow cords, cables, and tubes

References:

1. "Supplying Power to the Medical Industry: Retractable, Medical-Grade Power Cords." *AZoM.com*, Hunter Spring, 5 Dec. 2016, www.azom.com/article.aspx?ArticleID=13331.

2. "Support Arms & Tube Holders." *VBM*, www.vbm-medical.com/products/accessories-for-anaesthesia-intensive-care/support-arms-tube-holders/.

Photo credits:

Basket – from Colourbox.com

Bracket, recoiling mechanism, single channel – courtesy of Kimmy Ansems

Principle 72: Indicate radiation exposure

References:

1. "Relative Radioactivity Levels." *Berkeley RadWatch*, 2014, radwatch.berkeley.edu/dosenet/levels.

2. "How to Use a Geiger Counter." *Radiation Measurement Instruments*, ECOTEST, 8 Dec. 2016, ecotestgroup.com/press/blog/how-to-use-a-geiger-counter/.

3. Ropeik, David. "Fear of Radiation Is More Dangerous than Radiation Itself." *Aeon*, 21 May 2018, aeon.co/ideas/fear-of-radiation-is-more-dangerous-than-radiation-itself.

4. "Is Japan Reactor Crew Exposed to Fatal Radiation?" *National Geographic*, 19 Mar. 2011, news.nationalgeographic.com/news/2011/03/110317-japan-reactor-fukushima-nuclear-power-plant-radiation-exposure/.

5. Kennedy, Helen. "Japan's Fukushima Nuclear Plant on Fire; Radiation Levels Too High for Firefighters to Combat Blaze." *Daily News*, 15 Mar. 2011, www.nydailynews.com/news/world/japan-fukushima-nuclear-plant-fire-radiation-levels-high-firefighters-combat-blaze-article-1.120946.

6. Sylvester, Phil. "Japan Radiation: Are You Still at Risk from Fukushima?" *World Nomads*, 7 Sept. 2017, www.worldnomads.com/travel-safety/eastern-asia/japan/how-dangerous-is-the-radiation-in-japan.

7. Haffty, Bruce G., and Lynn D. Wilson. *Handbook of Radiation Oncology: Basic Principles and Clinical Protocols*. Jones and Bartlett Publishers, 2009.

8. Sarachick, Daniel. "Radiation Dosimeter Badge Usage." *Brown University*, http://www.brown.edu/Administration/EHS/public/badge_usage.pdf.

9. "Radiation Safety Solutions for Oil Fields." *Qal-Tek*, www.qaltek.com/oil-fields/.

10. "You're Being Exposed to Radiation -- But It's the Amount That Counts." *Los Angeles Times*, 15 Mar. 2011, articles.latimes.com/2011/mar/15/world/la-fg-radiation-comparison-20110315.

Principle 73: Shut off automatically

References:

1. "Getinge GEW 8668 Biotech/Laboratory Washer-Dryer." *Getinge Group*, ic.getinge.com/files/LS/complete-product-list/cleaning-and-decontamination-equipment/GEW%208668/Brochures/2812-gew-8668-washer-brochure-160321-en.pdf.

2. Golson, Jordan. "Rear-End Crashes Go Way Down When Cars Can Brake Themselves." *The Verge*, 27 Jan. 2016. https://www.theverge.com/2016/1/27/10854478/iihs-collision-warning-autobrake-volvo-city-safety-research.

Principle 74: Evacuate smoke

References:

1. Israel, Brett. "Inhaling Bacteria with Cigarette Smoke." *Scientific American*, Environmental Health News, 25 Nov. 2009, www.scientificamerican.com/article/cigarettes-smoking-bacteria-infection-pathogen/.

2. "Smoke-Free Operating Theatres." *BOWA-Electronic GmbH & Co. KG, Gomaringen*, Apr. 2015. https://bowa-medical.com/tradepro/shop/artikel/allgemein/BOWA-APG-MN031-644-SMOKE-EN-2015-04.pdf

3. "Why Don't Surgeons and OR Staff Protect Themselves from Plume?" *IC Medical - Global Leaders in Surgical Smoke Evacuation Technology*, 13 Oct. 2016, icmedical.com/dont-surgeons-staff-protect-plume/.

Principle 75: Protect against electric shock

References:

1. "Ground-Fault Circuit Interrupters (GFCI)." *Occupational Safety and Health Administration, United States Department of Labor*, www.osha.gov/SLTC/etools/construction/electrical_incidents/gfci.html.

2. "Preventing Electrical Shock with Ground Fault Circuit Interrupters." *Romitti Electric*, romitti.com/preventing-electrical-shock-with-ground-fault-circuit-interrupters/.

3. "The Importance of 'Grounding' Electrical Currents." *Platinum Electricians*, 6 Dec. 2017, www.platinumelectricians.com.au/blog/importance-grounding-electrical-currents/.

4. Sawyers, Harry. "Replacing Two-Prong Receptacles." *This Old House*, 16 Feb. 2018, www.thisoldhouse.com/ideas/replacing-two-prong-receptacles.

5. "GCSE Bitesize: Earthing and Double Insulation." *BBC*, www.bbc.co.uk/schools/gcsebitesize/science/edexcel_pre_2011/electricityworld/mainselectricityrev4.shtml.

6. "IEC 60417 — Graphical Symbols for Use on Equipment." *ISO - International Organization for Standardization*, IEC/SC 3C, 18 Feb. 2003, www.iso.org/obp/ui#iec:grs:60417:5172.

7. Reese, Charles D., and James Vernon Eidson. *Handbook of OSHA Construction Safety and Health*. CRC Press, 2006.

8. "Electrical Shock." *Encyclopædia Britannica*, Encyclopædia Britannica, Inc., 20 Apr. 2017, www.britannica.com/science/electrical-shock.

9. Giovinazzo, Paul. "The Fatal Current." *Elmwood Electric Inc.*, Feb. 1987, www.physics.ohio-state.edu/~p616/safety/fatal_current.html.

Photo credits:

Blow dryer electrical protection – Original graphic by Stefan Jellinek

Electric hand saw – from user: MdeVicente / Wikimedia Commons / CC0 1.0

Principle 76: Flash at an appropriate rate

References:

1. Howett, Gerald Leonard, et al. "Emergency Vehicle Warning Lights: State of the Art." U.S. Dept. of Commerce, National Bureau of Standards, 1978.

2. Rambler, Mark. "How They Put Motion in Motion Pictures." *The Washington Post*, WP Company, 10 Sept. 1997, www.washingtonpost.com/archive/1997/09/10/how-they-put-the-motion-in-motion-pictures/abae7c0e-dc66-4889-a52d-f25e363657b4/?utm_term=.1a1900b5fffb.

3. Clingan, Ian C. "Lighthouse." *Encyclopædia Britannica*, 9 Oct. 1998, www.britannica.com/technology/lighthouse.

4. Shafer, Patricia, and Sirven, Joseph. "Photosensitivity and Seizures." *Epilepsy Foundation*. 18 Nov. 2013. http://www.epilepsy.com/learn/triggers-seizures/photosensitivity-and-seizures.

5. "Car Turns Signals: Why They Blink, Make Sounds, and Look a Certain Way." University of Southern California, *Illumin*, illumin.usc.edu/286/car-turns-signals-why-they-blink-make-sounds-and-look-a-certain-way/.

Photo credits:

Fire alarm – from Max Pixel / CC0 Public Domain

Principle 77: Prevent glare and reflections

References:

1. Massey for the *Daily Mail*, Ray. "The Dazzling Sunsets That Kill 36 Drivers in 12 Months: Glare Contributes to 3,000 Accidents and Is Particularly Dangerous at This Time of Year." *Daily Mail Online*, Associated Newspapers, 16 Oct. 2013, www.dailymail.co.uk/news/article-2461972/Glare-sun-contributes-3-000-road-accidents-particularly-dangerous-time-year.html.

2. "Top Ski Goggles." *Top Ski Goggles Reviews and Advice*, topskigoggles.com/do-you-need-ski-goggles-top-5-reasons/.

3. "What to Do If Driving Into the Sun." *AAA Exchange*, exchange.aaa.com/safety/driving-advice/dangers-of-driving-into-sun/.

4. Morgan, Erinn. "Planning an Outdoor Adventure This Month? Make Sure You Have the Right Sunglasses." *All About Vision*, Feb. 2018, www.allaboutvision.com/sunglasses/polarized.htm.

5. United States, Federal Aviation Administration, Office of Aerospace Medicine, et al. July 2015. https://www.faa.gov/data_research/research/med_humanfacs/oamtechreports/2010s/media/201512.pdf

6. Sauter, Steven L., et al. "The Well-Being of Video Display Terminal Users: An Exploratory Study." U.S. Dept. of Health and Human Service. Center for Disease Control, 1983.

7. Simmons, Adam. "Matte vs Glossy." *PC Monitors*, 27 Dec. 16, pcmonitors.info/articles/matte-vs-glossy-monitors/.

8. Heiting, Gary. "Tired of Those Reflections on Your Eyeglasses in Photos?" *All About Vision*, www.allaboutvision.com/lenses/anti-reflective.htm.

Photo credits:

Boater – from Unsplash

Cockpit interior – used with permission from NuShield, Inc.

Driver – from Pexels

Principle 78: Guard against sudden static discharge

References:

1. "Vapor Self-Inspection Handbook." *Husky*, Husky Corporation, 2018, www.husky.com/technical-information/vapor-self-inspection-handbook/.

2. Will, Joanne. "Is Static Electricity at Pump a Real Danger?" *The Globe and Mail*, 11 May 2018, www.theglobeandmail.com/globe-drive/culture/commuting/is-static-electricity-at-pump-a-real-danger/article13367207/.

3. Biederman, Marcia. "Static Electricity Fires Are a Peril at the Pump." *The New York Times*, 27 July 2008, www.nytimes.com/2008/07/27/automobiles/27STATIC.html.

4. Hunter, Greg. "Can Cell Phones Cause Fires at Gas Pump?" *ABC News*, May 2017, abcnews.go.com/GMA/story?id=127836&page=1.

5. "Marines with the 26th MEU Perform Lift at Sunset on the Beach." *DVIDS*, 24 Oct. 2015, www.dvidshub.net/news/179831/marines-with-26th-meu-perform-lift-sunset-beach.

6. "Static Grounding Protection for Tank Trucks." *NewsonGale HOERBIGER Safety Solutions*, Newson Gale, 2016, www.newson-gale.com/faq-items/static-grounding-for-tank-trucks-22/.

7. Myers, Lisa, and Richard Gardella. "Warning: Scientists Say Gas Cans Carry Risk of Explosion." *NBCNews.com*, 2 Nov. 2015, www.nbcnews.com/news/world/warning-scientists-say-gas-cans-carry-risk-explosion-flna2D11691903.

Photo credits:

Principle 79: Add conspicuous warnings

References:

1. National Electrical Manufacturers Associations, editor. "ANSI Z535.4-2011 American National Standard Product Safety and Labels." American National Standard, 15 November, 2011, http://www.davis-inc.com/expert/docs/z535p4-2011.pdf.

2. Backinger, Cathy L., and Patricia A. Kingsley. *Write It Right: Recommendations for Developing User Instruction Manuals for Medical Devices Used in Home Health Care*. U.S. Dept. of Health and Human Services, Public Health Service, Food and Drug Administration, Center for Devices and Radiological Health, 1993.

3. NEMA. "ANSI Z535 Safety Alerting Standards." *NEMA*, www.nema.org/Standards/z535/Pages/default.aspx.

4. "ANSI Z535-6 – Product Safety Information in Product Manuals, Instructions, and Other Collateral Materials (2011)" *ANSI*, http://www.appliedsafety.com/wp-content/uploads/2011/08/ansi_z535dot6_article.pdf.

5. "Silently Guiding Safety: American National Standards for Safety Signs and Colors." *ANSI News and Publications*, 22 Sept. 2011, www.ansi.org/news_publications/news_story?menuid=7&articleid=724b7e4d-6e66-4a00-af99-6791da1e5018.

Principle 80: Prevent scalding

References:

1. "Scald Injury Prevention Educator's Guide." *American Burn Association: A Community Fire and Burn Prevention Program Supported by the United States Fire Administration Federal Emergency Management Agency*, Apr. 2017, ameriburn.org/wp-content/uploads/2017/04/scaldinjuryeducatorsguide.pdf.

2. Shields, Wendy C., et al. "Still Too Hot." *Journal of Burn Care & Research*, vol. 34, no. 2, 2013, pp. 281–287., doi:10.1097/bcr.0b013e31827e645f.

3. Weiman, Darryl S. "The McDonalds' Coffee Case." *The Huffington Post*, 7 Jan. 2017, https://www.huffingtonpost.com/darryl-s-weiman-md-jd/the-mcdonalds-coffee-case_b_14002362.html

4. Brown, Fredericka and Kenneth R. Diller. "Calculating the Optimum Temperature for Serving Hot Beverages." *Burns* 34 (2008): 648-654.

5. "Scald Safety." *Shriners Hospitals for Children*, https://www.shrinershospitalsforchildren.org/shc/scald-safety.

Exemplar 8: Steam iron

Photo credits:

Principle 81: Enable safety feature testing

References:

1. "NEMA and UL Announce Revisions to UL 943 GFCI Standard." *Electrical Line Magazine*, 29 Jan. 2015, electricalline.com/nema-and-ul-announce-revisions-ul-943-gfci-standard.

2. Rouse, Margaret. "What Is POST (Power-On Self-Test)?" *WhatIs.com*, Apr. 2005, whatis.techtarget.com/definition/POST-Power-On-Self-Test.

3. "What Is an LVAD? How Does It Work?" *MyLVAD*, https://www.mylvad.com/content/what-lvad-how-does-it-work.

4. "HeartWare® Ventricular Assist System Instructions for Use." *HeartWare*, 2012. http://www.heartware.com/sites/default/files/uploads/docs/ifu00001_rev_15.pdf.

Photo credits:

GFCI receptacle – courtesy of Rachel Aronchick

Left Ventricular Assist Device – Reproduced with permission of Medtronic, Inc.

Principle 82: Add shape-coding

References:

1. Hou, Ming, Simon Banbury, and Catherine Burns. *Intelligent Adaptive Systems: An Interaction-Centered Design Perspective*. CRC Press, Taylor & Francis Group, 2017.

2. Rassias, Athos J, et al. "A Prospective Study of Tracheopulmonary Complications Associated with the Placement of Narrow-Bore Enteral Feeding Tubes." *Crit Care*, vol. 2, no. 1, 1998, pp. 25–28., doi:10.1186/cc120.

3. Pew, Richard. "Alphonse Chapanis: Pioneer in the Application of Psychology to Engineering Design." *Association for Psychological Science*, Apr. 2010, https://www.psychologicalscience.org/observer/alphonse-chapanis-pioneer-in-the-application-of-psychology-to-engineering-design.

Photo credits:

Airplane wheels – from Colourbox.com

B-17 Airplane – from Wikimedia Commons / Public Domain

Wheel lever – used with permission from SKALARKI Electronics

Principle 83: Provide backup power

References:

1. "What Is Transfer Time?" *Sunpower UK*, 1 July 2014, www.sunpower-uk.com/glossary/what-is-transfer-time/.

2. Davies, Alex. "Why Volcanic Ash Is So Terrible for Airplanes." *Wired, Conde Nast*, 3 June 2017, www.wired.com/2014/08/volcano-ash-planes/.

3. Crivelli, Davide. "What Happens When a Bird Strikes a Plane?" *The Conversation*, 25 Sept. 2017, theconversation.com/what-happens-when-a-bird-strikes-a-plane-84502.

4. Dismukes, Kim. "Hydraulic System." *NASA*, 7 Apr. 2002, spaceflight.nasa.gov/shuttle/reference/shutref/orbiter/hyd/.

Photo credits:

Aircraft – from user: Curimedia / Wikimedia Commons / CC-BY-2.0

Sailboat – courtesy of Veerle Migchelbrink

Ski lift – from user: Crystalmountainskier / Wikipedia / CC-BY-SA-3.0

Ski lift motor – from Pxhere / Public Domain

Spacecraft – from Wikimedia Commons / Public Domain

Spacecraft motor – from Wikimedia Commons / Public Domain

Principle 84: Don't depend solely on color

References:

1. Rathus, Spencer. *Psychology: Concepts and Connections*. 10th ed., Wadsworth CENGAGE Learning, 2011.

2. "Facts about Color Blindness." *National Eye Institute*, U.S. Department of Health and Human Services, 1 Feb. 2015, https://nei.nih.gov/health/color_blindness/facts_about.

3. Oliveira, Izautino P., et al. "A Vision of Traffic Lights for Color-Blind People." *SMART 2015 : The Fourth International Conference on Smart Systems*, Devices and Technologies, 2015. https://www.thinkmind.org/download.php?articleid=smart_2015_2_40_40071.

4. "What Is Colorblindness and the Different Types?" *Waggoner*, www.colorvisiontesting.com/color2.htm.

5. "Color Blindness." *Color Blindness*, www.colour-blindness.com/.

6. Subrahmanyam, M, and S Mohan. "Safety Features in Anaesthesia Machine." *Indian Journal of Anaesthesia,* vol. 57, no. 5, 2013, pp. 472–480., doi:10.4103/0019-5049.120143.

7. Weinger, Matthew B., et al. *Handbook of Human Factors in Medical Device Design*. CRC Press/Taylor & Francis, 2010.

8. "What Kind of Colorblind Are You?" *The Huffington Post*, TheHuffingtonPost.com, 6 Sept. 2013, www.huffingtonpost.com/2013/09/06/colorblind-quiz_n_3867952.html.

9. Tsukayama, Hayley. "10 Surprising Facts about Mark Zuckerberg." *The Washington Post*, WP Company, 30 May 2012, www.washingtonpost.com/business/technology/10-surprising-facts-about-mark-zuckerberg/2012/05/30/gJQAJ9yJ2U_story.html?utm_term=.e3662de3f18a.

Photo credits:

Ishihara plate – from Wikimedia Commons / Public Domain

Traffic light – from Wikimedia Commons / Public Domain

Principle 85: Enable emergency calls

References:

1. "LIFE ALERT Official Website - I've Fallen and I Can't Get up!®." *LIFE ALERT Official Website - I've Fallen and I Can't Get up!®*, www.lifealert.com/.

2. "Voice Radio Communications Guide for the Fire Service." *U.S. Fire Administration*, FEMA, June 2016.https://www.usfa.fema.gov/downloads/pdf/publications/voice_radio_communications_guide_for_the_fire_service.pdf

3. Menno, Christian. "Emergency Call Boxes along the PA Turnpike Being Removed as Cellphone Usage Rises." *Burlington County Times*, 15 Sept. 2017, www.burlingtoncountytimes.com/64ef6b0e-f151-5d85-9a7c-75f355e9775e.html.

4. "What Does Mayday Mean?" *Wonderopolis*, wonderopolis.org/wonder/what-does-mayday-mean.

5. Kahle, Laurie. "Breitling's Emergency II Doubles Down on Rescue Technology." *Forbes Magazine*, 31 July 2013, www.forbes.com/sites/lauriekahle/2013/07/31/breitlings-emergency-ii-doubles-down-on-rescue-technology/#697a631aa3f5.

6. "What Is a Personal Locator Beacon (PLB)?" *Marine Rescue Technologies*, www.mrtsos.com/plbs-explained/what-is-a-personal-locator-beacon-plb.

Photo credits:

Emergency SOS screen – courtesy of Jon Tilliss

Helicopter – from U.S. Indo-Pacific Command by Sgt. Daniel K. Johnson / Public Domain

Smartphone – from Pixabay / Public Domain

Principle 86: Use graphical instructions

References:

1."13 Reasons Why Your Brain Craves Infographics." *NeoMam Studios*, neomam.com/interactive/13reasons/.

2. Trafton, Anne. "In the Blink of an Eye." *MIT News*, 16 Jan. 2014, news.mit.edu/2014/in-the-blink-of-an-eye-0116.

3. Whitehouse, Andrew J.O., et al. "The Development of the Picture-Superiority Effect." *British Journal of Developmental Psychology*, vol. 24, no. 4, Nov. 2006, pp. 767–773., doi:10.1348/026151005X74153.

4. Tate, Andrew. "10 Scientific Reasons People Are Wired to Respond to Your Visual Marketing: Learn." *Canva*, 17 May 2018, www.canva.com/learn/visual-marketing/.

5. Levie, Howard W., and Richard Lentz. "Effects of Text Illustrations: A Review of Research." *Educational Technology Research and Development*, vol. 30, no. 4, Dec. 1982, pp. 195–232., link.springer.com/article/10.1007/BF02765184.

Principle 87: Account for untrained use

References:

1. "Forklift Training Certification." *Safety Council of Palm Beach County, Inc.*, 2014, http://safetycouncilpbc.org/forklift-training/

2. "How do I Become a Commercial Airline Pilot?" *Pilot Career News*, 2018, https://www.pilotcareernews.com/how-to-be-a-pilot/.

3. Bisantz, Ann M., et al. *Cognitive Systems Engineering in Health Care.* CRC Press LLC, 2015.

4. APSF Committee on Technology. "Training Anesthesia Professionals to Use Advanced Medical Technology." *Anesthesia Patient Safety Foundation*, 2018, https://www.apsf.org/article/training-anesthesia-professionals-to-use-advanced-medical-technology/.

5. Inspired by Mayer, Caroline E., "Why Won't We Read the Manual?" *The Washington Post*, 26 May 2002, https://www.washingtonpost.com/archive/business/2002/05/26/why-wont-we-read-the-manual/b7f08098-1d08-4d67-9e3e-8f3814f4d90a/?utm_term=.cbf06779a479

6. Burgstahler, Sheryl, Ph.D. "Universal Design of Instruction (UDI): Definition, Principles, Guidelines, and Examples." *DO-IT*, University of Washington (UW), 2018, http://www.washington.edu/doit/universal-design-instruction-udi-definition-principles-guidelines-and-examples.

Photo credits:

Forklift, anesthesiologist – from Colourbox.com

Car, tractor, grill scene – courtesy of Cory Costantino

Principle 88: Provide guidelines

References:

1. Zimmer, Fred. "Preventing Engine Ingestion Injuries When Working Near Airplanes." *Aeromagazine*, www.boeing.com/commercial/aeromagazine/articles/qtr_3_08/pdfs/AERO_Q308_article4.pdf.

Photo credits:

Jet nacelle – courtesy of Jon Tilliss

Miter saw – from iStock

Train station platform, car odometer, car backup camera – courtesy of Rachel Aronchick

Principle 89: Use voice prompts

References:

1. "Instructions for Use AED Trainer." *Cardiac Science Corporation*, 2007, www.cardiacscience.com/cardiac-assets/manuals/5281.pdf.

2. "Introduction to TCAS II - Version 7.1." *U.S. Department of Transportation*, 28 Feb. 2011, www.faa.gov/documentLibrary/media/Advisory_Circular/TCAS%20II%20V7.1%20Intro%20booklet.pdf.

3. "What Will I See and Hear from Nest Protect?" *Nest*, Nest Labs, Inc., nest.com/support/article/What-will-I-see-and-hear-from-Nest-Protect.

4. Edworthy, J., et al. "The Use of Male or Female Voices in Warnings Systems : A Question of Acoustics." *Noise Health*, vol. 6, no. 21, 2003, pp. 39–50., www.noiseandhealth.org/article.asp?issn=1463-1741;year=2003;volume=6;issue=21;spage=39;epage=50;aulast=Edworthy.

5. Green, Paul, et al. "Suggested Human Factors Design Guidelines for Driver Information Systems." *The University of Michigan Transportation Research Institute*, Office of Safety and Traffic Operations R&D, Nov. 1993, http://citeseerx.ist.psu.edu/viewdoc/download?doi=10.1.1.588.185&rep=rep1&type=pdf.

6. Weinger, Matthew Bret, Michael E. Wiklund, and Daryle Jean Gardner-Bonneau. *Handbook of Human Factors in Medical Device Design.* CRC Press, 2010.

Photo credits:

AED – courtesy of Kimmy Ansems

Principle 90: Enable fast action in an emergency

References:

1. Bonsor, Kevin. "How Ejection Seats Work." *HowStuffWorks*, 27 June 2001, science.howstuffworks.com/transport/flight/modern/ejection-seat3.htm.

2. "5 Cool Pieces of Skydiving Gear You Didn't Know About." *Skydive Long Island*, 16 Jan. 2007, www.skydivelongisland.com/about/articles/5-cool-pieces-of-skydiving-gear-you-didn-t-know-about/.

3. "Use Emergency SOS on Your iPhone." *Apple Support*, 2 Nov. 2017, support.apple.com/en-us/HT208076.

Photo credits:

Air stewardess at emergency exit – courtesy of Donya Bamford

iPhone – from user: 100 lion / Wikimedia Commons / CC-BY-SA-4.0

Jet ejection seat – Matilda Ahl / Flygvapenmuseum (Swedish Air Force Museum) / CC-BY-SA-4.0

Exemplar 9: AED

Photo credits:

Defibrillation – from user: Rama / Wikimedia Commons / CC-BY-SA-2.0

Principle 91: Make it touch free

References:

1. Nichols, Sean K. "Touchless Technology Helps Facilitate Infection Prevention Best Practices." *Infection Control Today*, 1 Sept. 2007, www.infectioncontroltoday.com/personal-protective-equipment/touchless-technology-helps-facilitate-infection-prevention-best.

2. Keiran, Monique. "Monique Keiran: Germs Thrive with Touchless Technology." *Times Colonist,* 20 Nov. 2016, www.timescolonist.com/opinion/columnists/monique-keiran-germs-thrive-with-touchless-technology-1.2939619.

3. Askar, Nadia. "A Hands-Free, Germ-Free Future: A Look into the Offerings of Residential Touchless Faucets." *Plumbing & Mechanical*, vol. 32, no. 8, Oct. 2014, www.cinaton.com/file/PM_Mag_A_hands-free_germ-free_future.pdf.

Photo credits:

Toilet, kitchen sink, bathroom sink – courtesy of Cory Costantino

Steering wheel – courtesy of Jon Tilliss

Principle 92: Guard against over inflation

References:

1. Hefny, Ashraf F., et al. "Blast Injuries of Large Tyres: Case Series." *International Journal of Surgery*, vol. 8, 2010, pp. 151–154., www.journal-surgery.net/article/S1743-9191(09)00178-2/pdf.

2. Purwar, Shreyansh Kumar. "Automatic Tire Inflation System." *International Research Journal of Engineering and Technology (IRJET),* vol. 4, no. 4, Apr. 2017, www.irjet.net/archives/V4/i4/IRJET-V4I4489.pdf.

3. Prindle, Drew. "Know Your Tire Pressure at a Glance with This Clever Valve Stem Cap." *Digital Trends*, 19 Feb. 2014, www.digitaltrends.com/cool-tech/know-tire-pressure-glance-clever-valve-stem-cap/.

4. "Ventilator/Ventilator Support." U.S. National Library of Medicine, *PubMed Health*, 11 June 2014, www.ncbi.nlm.nih.gov/pubmedhealth/PMH0063006/.

5. Safety Is by Design at I.C. Medical Company. *IC Medical*, 1 May 2017, www.icmedical.com/safety-design-c-medical-company/.

6. Gurgle, Amy S. "Overinflated Exercise Balls Pose Risk of Injuries." *Regan Zambri Long*, 23 April, 2009, https://www.rhllaw.com/2009/04/23/overinflated-exercise-balls-pose-risk-of-injurie/.

7. "Fitness Balls Recalled by EB Brands Due to Fall Hazard; New Assembly Instructions Provided." *U.S. Consumer Product Safety Commission*, 29 May 2017, www.cpsc.gov/Recalls/2009/fitness-balls-recalled-by-eb-brands-due-to-fall-hazard-new-assembly-instructions.

8. "Exercise Ball User Manual US." 2014, www.aldi.us/fileadmin/fm-dam/Services/warranties/warranties_Jan/1.1.14_Crane_Exercise_Ball_Manual.pdf.

Principle 93: Label toxic substances

References:

1. "Federal Hazardous Substances Act (FHSA) Requirements." *U.S. Consumer Product Safety Commission*, https://www.cpsc.gov/Business--Manufacturing/Business-Education/Business-Guidance/FHSA-Requirements .

2. 2016 TEI Expert - June. "Warnings for Hazardous Chemical Products: When Are They Enough?" *The Expert Institute*, 6 June 2017, www.theexpertinstitute.com/warnings-for-hazardous-chemical-products-when-are-they-enough/.

3. Black , Jim. "Be Wary of Household Hazards." *Herald-Standard*, 12 Jan. 2015, www.heraldstandard.com/healthy_living/be-wary-of-household-hazards/article_3c588017-f5ef-527c-834f-61544297667c.html.

4. "Peel Adhesion Bond Strength of Labels." *TestResources*, www.testresources.net/applications/testtypes/peel-test/peel-adhesion-bond-strength-of-labels/.

5. "FirstDefender™ RMX Handheld Chemical Identification." *Thermo Fisher Scientific*, www.thermofisher.com/order/catalog/product/FIRSTDEFENDERRMX?ICID=search-product%2F.

6. "National Fire Protection Association Hazard Identification System." *American Chemical Society*, 27 Mar. 2017, www.acs.org/content/acs/en/chemical-safety/guidelines-for-chemical-laboratory-safety/resources-supporting-guidelines-for-chemical-laboratory-safety/national-fire-protection-association-hazard-identification.html.

7. "Hazmat Placard Specifications and Requirements." *LabelMaster*, 14 Jan. 2014, www.labelmaster.com/hazmat-source/hazmat-placard-specifications.

Principle 94: Include brakes

References:

1. Reuters. "Runaway Beverage Cart on American Airlines Flight Gave Me Brain Damage: Lawsuit." *New York Post*, 16 June 2017, nypost.com/2017/06/16/runaway-beverage-cart-on-american-airlines-flight-gave-me-brain-damage-lawsuit/.

2 ."Brake System Evolution: A History." *Greg Monforton and Partners Injury Lawyers*, 2018, https://www.gregmonforton.com/evolution-brake-systems.html.

Principle 95: Provide restraints

References:

1. Harris, Donna Lawrence. "1964 Brought Seat Belts for Every Buyer." *Automotive News*, 26 June 1996.

2. "A History of Seat Belts." *DefensiveDriving.com*, 14 Sept. 2016, www.defensivedriving.com/blog/a-history-of-seat-belts/.

3. Myers, Alexis. "Washington, Other States Eye School Bus Seat Belts." *USA Today*, 25 Jan. 2017, www.usatoday.com/story/money/cars/2017/01/25/school-bus-seat-belts/97063570/.

4. Tracy, Todd. "Seat Belt-Submarining Can Cause Occupants Ejection during an Accident." *The Tracy Law Firm*, 26 Nov. 2012, www.vehiclesafetyfirm.com/blog/general/seat-belt-submarining-violets-preventing-occupants-ejection-causing-it-to-be-a-crashworthiness-case/

5. "Fatalities." *Safercar.gov*, NHTSA, www.safercar.gov/Vehicle-Shoppers/Rollover/Fatalities.

6. Zennie, Michael. "Laughing in the Face of Death: Workers Pictured Fooling around as They Built Some of America's Most Iconic Buildings: Where One Person Died for Every $1m Spent." *DailyMail.com*, 21 Jan. 2015, www.dailymail.co.uk/news/article-2920453/Laughing-face-death-Incredible-pictures-construction-workers-fooling-built-America-s-iconic-buildings.html.

7. OSHA, "Construction Safety and Health: 'Fall' Hazards Trainer Guide." US Department of Labor, https://www.osha.gov/dte/grant_materials/fy07/sh-16586-07/1_fall_hazards_trainer_guide.pdf

8. "High-Tech Vehicle Safety Systems." *CARSP | Seat Belt Pretensioners | Canadian Association of Road Safety Professionals*, www.carsp.ca/research/resources/high-tech-vehicle-safety-systems/seat-belt-pretensioners/.

Photo credits:

Construction worker – from user: Western Area Power / Flickr / CC-BY-2.0

Patient restraint – James Heilman, MD / CC-BY-SA-4.0

Principle 96: Make PPE available and usable

References:

1. Wilson, Charles. "Does Safety Equipment Really Work?" *Charles Wilson Engineers Ltd*, 5 Sept. 2017, www.cwplant.co.uk/safety-equipment-really-work/

2. European Lung White Book. "Occupational Lung Diseases." *Occupational Lung Diseases*, www.erswhitebook.org/chapters/occupational-lung-diseases/.

3. Groce, Donald F. "Drive Home the Value of Gloves: Hand Injuries Send a Million Workers to ERs Each Year." Industrial Safety & Hygiene News, ISHN, 6 Sept. 2012.

4. Workplace Safety North, A Health & Safety Ontario Partner. "Health and Safety Resources." Sept. 2003.https://www.workplacesafetynorth.ca/sites/default/files/Hand%20Injuries%20Participant%20Manual.pdf

5. "Understanding NFPA 1710 Response Times." PURVIS Public Safety, *Purvis Systems Public Safety Division*, 18 Oct. 2014, purvispublicsafety.com/2014/10/18/nfpa-1710-response-times/.

Photo credits:

Dispensing system – courtesy of Scott Coleman

Integrated helmet – CC-SA-1.0

Protective gloves – from Pixabay / Public Domain

Stacked goggles, stacked helmets – courtesy of Brenda van Geel

Toolbelt – Magnus Mertens / CC-BY-SA-2.0

Principle 97: Slow down falling objects

References:

1. Sharkey, Joe. "Reinventing the Suitcase by Adding the Wheel." *The New York Times*, 4 Oct. 2010, www.nytimes.com/2010/10/05/business/05road.html.

2. The Associated Press. "Robert Kearns, 77, Inventor of Intermittent Wipers, Dies." *New York Times*, 26 Feb. 2005, www.nytimes.com/2005/02/26/obituaries/robert-kearns-77-inventor-of-intermittent-wipers-dies.html.

3. Glass, Allison S., et al. "No Small Slam: Increasing Incidents of Genitourinary Injury from Toilets and Toilet Seats." *BJU International*, vol. 112, no. 3, 2013, pp. 398–403., doi:10.1111/bju.12173.

4. "CPSC Adopts Voluntary Standard for Toy Chests." *United States Consumer Product Safety Commission*, 18 Aug. 1983, www.cpsc.gov/content/cpsc-adopts-voluntary-standard-for-toy-chests.

5. United States, Consumer Product Safety Commission. "Standard Consumer Safety Specification for Toy Safety." Designation: F963-17, ASTM International, 2017.

Photo credits:

Self-closing tailgate, kitchen drawer – courtesy of Rachel Aronchick

Truck tailgate – used with permission from Dee Zee

Principle 98: Eliminate pinch points

References:

1. "Crush Injury." *Injury Information.com,* Newsome Law Firm, www.injuryinformation.com/injuries/crush-injury.php.

2. "Why Soft Close Drawers and Cabinets." *Cabinet Hardware, Inc.,* 18 Jan. 2018, www.cabinethdw.com/soft-close-drawers-cabinets/.

3. Davis, Clint. "29,400 gb Qbit Strollers Recalled after Injuries Include Broken Wrist, Elbow and Stitches." *NBC26,* Scripps Media, Inc., 21 Dec. 2016, www.nbc26.com/news/national/29400-qbit-strollers-recalled-after-injuries-include-broken-wrist-elbow-and-stitches.

Photo credits:

Air bag lift jack, child's fingers on drawer, stroller – from Colourbox.com

Bicycle guard – from Richard Masoner / Flickr / CC BY-SA-2.0

Paper cutter, scissor lift – courtesy of Ruben Post

Principle 99: Enable sterilization

References:

1. Ghose, Tia. "3.5-Billion-Year-Old Fossil Microbial Community Found." *LiveScience,* Purch, 13 Nov. 2013, www.livescience.com/41191-ancient-microbe-fossils-found.html.

2. May, Kate Torgovnick. "6 Great Things Microbes Do for Us." *TedBlog,* 10 July 2012, blog.ted.com/6-great-things-microbes-do-for-us/.

3. "National Center for Health Statistics." *Centers for Disease Control and Prevention,* 20 Jan. 2017, www.cdc.gov/nchs/fastats/pneumonia.htm.

4. Epidemiology and Statistics Unit Research and Health Education Division. "Trends in Pneumonia and Influenza Morbidity and Mortality." *American Lung Association,* Nov. 2015, www.lung.org/assets/documents/research/pi-trend-report.pdf.

5. Pappas, Stephanie. "The Frankenstein of Giant Viruses Found in Sewage Plant." *LiveScience,* 7 Apr. 2017, www.livescience.com/58586-frankenstein-giant-virus-found-sewage-plant.html.

6. "What Is the Best Sterilization Method for Your Lab?" *Laboratory-Equipment.com,* 24 Mar. 2016, www.laboratory-equipment.com/blog/all-laboratory-equipment-blogs/best-sterilization-method-lab/.

7. "Largest Autoclave." *Guinness World Records,* 2006, www.guinnessworldrecords.com/world-records/largest-autoclave-.

8. Morisseau, D., et al. "New Applications for Accelerators in Pharmaceutical Processes." *International Atomic Energy Agency,* www-pub.iaea.org/MTCD/publications/PDF/P1433_CD/datasets/papers/sm_eb-08.pdf.

9. Tarawneh, Walid. "Dialysis Water Pre-Treatment Plant." *LinkedIn,* 21 Oct. 2017, www.linkedin.com/pulse/dialysis-water-pre-treatment-plant-dr-eng-walid-tarawneh.

10. Hemmerich, Karl J. "Polymer Materials Selection for Radiation-Sterilized Products." *MDDI Online,* 1 Feb. 2000, www.mddionline.com/article/polymer-materials-selection-radiation-sterilized-products/.

11. "Anatomy of an Ethylene Oxide Sterilization Process | TechTip." *STERIS AST,* 2017, www.steris-ast.com/tech-tip/anatomy-ethylene-oxide-sterilization-process/.

Photo credits:

Autoclave – from user: SystecAutoclaves / Wikimedia Commons / CC-BY-SA-4.0

Tie – from Peakpx / CC0 Public Domain

Principle 100: Minimize repetitive motion

References:

1. "Repetitive Motion Injury." *Johns Hopkins Medicine,* www.hopkinsmedicine.org/healthlibrary/conditions/physical_medicine_and_rehabilitation/repetitive_motion_injury_85,P01176/.

2. "Thoracic Outlet Syndrome." *Mayo Clinic,* Mayo Foundation for Medical Education and Research, 27 Aug. 2016, www.mayoclinic.org/diseases-conditions/thoracic-outlet-syndrome/symptoms-causes/syc-20353988.

3. Ma, Benjamin, MD. "Impingement Syndrome." *U.S. National Library of Medicine,* MedlinePlus, 27 Sept. 2017, medlineplus.gov/ency/imagepages/19614.htm.

4. "Lateral and Medial Epicondylitis." *Summit Orthopedics,* 2018, www.summitortho.com/services/elbow/lateral-and-medial-epicondylitis/.

5. Newman, Tim. "Ganglion Cyst: Symptoms, Causes, and Treatment." *Medical News Today,* 21 Nov. 2017, www.medicalnewstoday.com/articles/156995.php.

6. Shen, Shixin (Cindy), and Ronald A. House. "Hand-Arm Vibration Syndrome." *Canadian Family Physician*, vol. 63, no. 3, Mar. 2017, pp. 206–210., www.ncbi.nlm.nih.gov/pmc/articles/PMC5349719/.

7. Smith, Korydon H., and Wolfgang F. E. Preiser. *Universal Design Handbook*. Second ed., McGraw-Hill, 2011

8. "The Ageing Workforce and MSDs." *Fit for Work*, 15 June 2016, fitforwork.org/blog/the-ageing-workforce-and-musculoskeletal-disorders-msds/.

9. Elston, Laura. "Shorter, Tougher Life for Left-Handers." *The Independent*, 12 Sept. 2000, www.independent.co.uk/student/postgraduate/mbas-guide/shorter-tougher-life-for-left-handers-5369601.html.

10. Mannonen, Sari, et al. "The Benefits of Electronic Pipetting: How to Choose the Correct Pipettor?" *International Labmate*, 2002, www.vivaproducts.com/downloads/eline-electronic-pipette-benefits.pdf.

Photo credits:

Anatomy of the human body – from Piotr Siedlecki / Public Domain Pictures / Public Domain

Pipette – from Pxhere / Public Domain

Exemplar 10: Stretcher
Photo credits:

Emergency personnel with stretcher – from Colourbox.com

About the authors

Michael Wiklund is an internationally recognized human factors engineering expert with more than 30 years of experience. He has authored several books about designing products for safe, effective, and satisfying use. He serves as general manager of the human factors research and design practice at UL (Underwriters Laboratories), is a professor of the practice at Tufts University, and frequently speaks at industry events focused on safety. Michael is an aspiring painter who focuses on architectural and mechanical subjects.

Kimmy Ansems holds her master's and bachelor's in industrial design from the University of Technology in Eindhoven. She has been practicing human factors engineering for nearly four years within the domain of medical technology. Her addiction to design can also be found at home, where she collects various space age design treasures.

Rachel Aronchick is a certified human factors professional with a master's in digital media and interactive design from Northeastern University and a bachelor's in human factors engineering from Tufts University. Rachel has been practicing human factors engineering for more than five years, primarily with a focus on making medical devices safe and usable. In her spare time, she enjoys putting her soccer cleats to use and braving the cold for some time on the ski slopes.

Cory Costantino is a certified human factors professional and holds a master's in human factors in information design from Bentley University. He has taught numerous courses within the field of design at Wentworth Institute of Technology. Cory has been practicing design for almost 20 years in consulting, corporate, and start-up companies across medical and consumer product domains. Cory also holds a sixth-degree black belt in the martial art of Tang Soo Do.

Alix Dorfman holds her master's in human factors and applied cognition from George Mason University and her bachelor's in psychology and economics from Cornell University. She has been practicing human factors engineering for more than five years within the domains of military and medical technology, with a current focus on exoskeletons. Aside from writing about design, she enjoys writing music on her guitar and exploring the power of biomechanics firsthand as an enthusiastic athlete.

Brenda van Geel holds her master's in design for interaction and her bachelor's in architecture from Delft University of Technology. She has been practicing human factors engineering and user experience design for three years within the medical technology domain. Brenda is training to become a Fosbury flop expert (i.e., jumping style used for the track and field high jump event).

Jonathan Kendler is a user interface designer and human factors engineer with more than 20 years of experience. He has designed user interfaces for various safety-critical products, including dialysis machines, robotic surgical systems, and infusion pumps. In addition to co-authoring articles and books on design and human factors engineering, Jonathan has published children's books and poems under a pseudonym.

Valerie Ng holds her master's in fine arts and her human-computer interaction certificate from Tufts University. She has been practicing user experience and user interface design for more than four years within the medical field. To balance the focus on safety in her line of work, Val got her motorcycle license and plans to ride through the Badlands someday.

Ruben Post holds his PhD in industrial design engineering from Delft University of Technology and is the editor-in-chief of <u>The Magazine for Human Factors</u> in The Netherlands. He has also taught usability, product evaluation, and product perception courses at Delft University of Technology. Ruben has been practicing human factors engineering for three years within the domain of medical device usability. When Ruben isn't studying product interaction and sensory perception, he enjoys personally experiencing both while driving his motorcycles.

Jon Tilliss is a certified human factors professional with a master's in digital media and interactive design and a strong foundation in user experience research. He is a part-time lecturer at Tufts University, where he teaches a course on user interface design. Jon has more than 10 years of experience leading cross-functional teams to design and deliver safe and usable solutions that delight users in the medical, healthcare, and telecommunications domains. In his free time, Jon can be found cuddling with his dogs, watching the Denver Broncos, and making puns.

T - #0204 - 230425 - C278 - 279/216/15 [17] - CB - 9780367188313 - Matt Lamination